Aftermath

John Schofield

Aftermath

Readings in the Archaeology
of Recent Conflict

 Springer

John Schofield
English Heritage
Swindon SN2 2GZ
United Kingdom
john.schofield@english-heritage.org.uk

ISBN: 978-0-387-09464-9   e-ISBN: 978-0-387-88521-6
DOI: 10.1007/978-0-387-88521-6

Library of Congress Control Number: 2008940908

Printed on acid-free paper

springer.com

Contents

For My Mother

List of Figures

List of Tables

Acknowledgements

For each of the 14 chapters in this book there are people to be thanked, and credits and acknowledgements owed. However, as all of these people are fully acknowledged in the original chapters I have chosen not to repeat the information here. I have chosen instead to include a generic acknowledgement and to credit those who have contributed most to the overall direction my work has taken over the past decade, and those therefore who have helped shape the present volume and the ideas that it contains. On the one hand are those professional colleagues who have provided such stimulating and pleasant company over the years, in addition to their contribution to shaping many of the conceptual frameworks that are given concrete expression amongst these pages: members of English Heritage's Military and Naval Strategy Group for example, who meet three times every year to provide update on progress and prospects in recent military matters; members of English Heritage's Characterisation Team whose collective ability to see the bigger picture, to see the wood for the trees, has been a constant and refreshing presence since the team's formation in 2001; and those consultants, curators, contractors and other specialists who have undertaken so much valuable work, some at the behest of English Heritage, and some independently, work that has truly taken the subject forward in leaps and bounds since the mid 1990s. All have always been willing to share their results, and to discuss them openly and with enthusiasm. The academic rigour displayed by these researchers, almost without exception, alongside their commitment and their generosity, has been a constant source of inspiration.

Military archaeology – like industrial archaeology and other specialisms I suspect – provides a stimulating environment within which to work: the great enthusiasm and expertise of its proponents allows the subject to progress as it does, with new ideas and perspectives, and new data. It has been a personal highlight of my professional career to share this intellectual ground with some of the most committed and interesting people, whose industry and enthusiasm will hopefully continue for many years yet. The fact that some colleagues are nearing the end of their professional careers as others are just beginning is a most hopeful sign of what is to come.

More specifically I am grateful to those at Springer who encouraged this publication, and have guided it through to completion. Teresa Krauss deserves particular mention for her insightful and intelligent suggestions, and her enthusiastic encouragement of the closing chapter about which, I must confess, I had my doubts!

I also owe a huge personal debt to those amongst my friends and family that have tolerated my enthusiasm over the years, and endured the numerous pillbox walks that have become such a feature of holidays and journeys. Hopefully, this collection provides at least part of an answer to the question they have often asked, and never received a satisfactory answer to: 'so what sort of archaeology do you do exactly?' I have previously noted (in Matériel Culture) my father's influence on my life and work, and in Combat Archaeology the debt to my two older children. Here I want to acknowledge in particular the continuing support and interest shown by my mother in all the work I have undertaken, including that represented in this particular collection. Other than me, my mother is probably the only person to have read everything I have published. Hopefully, she won't now sit down to read this book from cover to cover, having previously read all the chapters in their original form. For the support shown throughout my professional career, this collection is dedicated to her, with as much love, respect and gratitude as I can muster.

15 April 2008   Oulu, Finland

Preface

The archaeology of recent conflict has emerged as a credible, popular and significant field of archaeological endeavour and heritage concern in the past decade, building on previous work by, amongst others, the late Henry Wills and Andrew Saunders. I should be quite clear about this from the outset: I make no claim to have initiated this development, or driven it in any sense. I have merely contributed, with others, at a time of progress and change, witnessing and participating in a movement that caused what was previously a fringe (and largely amateur) pursuit to become serious and worthy; professional indeed. Admittedly my role, directing – or to be precise co-ordinating – much of the work undertaken or commissioned by English Heritage (the state heritage agency in England, and advisor to the British government on cultural heritage matters), has put me in a somewhat privileged position, the evidence of which is clear in some of the pages that follow. This book comprises my own work (shared where co-authored) but sometimes only in as much as the words and the perspectives represented are of my construction. Some of the projects that these chapters draw upon are the works of others, and that is fully acknowledged if not here then in the original publications.

It is my hope that this book will contribute to the continuation of this trend, and perhaps even to establishing further, presently unforeseen, research directions, particularly by those beginning their studies, or early in their professional careers. If this book inspires even a handful of students to undertake dissertation topics into some aspect of recent conflict archaeology, or an early-career heritage professional to undertake or commission a research project that may not otherwise have been conceived, then it will have been a success. The main purpose of this book is to make a series of texts more accessible for students interested in this area of work, texts which together chart the development and maturation of this area of research in the period 1995–2007. While some of the texts included here are easily obtained in their original form, others were originally published overseas in more obscure and inaccessible places. It is hoped that publishing them again, and together, will produce the basis for further thought and research endeavour. It is this collection as a whole, and the ideas and the examples it contains, that will hopefully generate a few sparks of interest.

A Note on Currency

In re-working these previously published chapters, I have chosen not to fully research the current situation where a particular management or conceptual regime existed at the time of writing. In a few cases (Dora in Chap. 1, District Six in Chap. 2) one or two notes are included that outline the changes to have occurred in the intervening period, but typically that is not the case. In most chapters, there is the potential for a situation presented as current to have been reviewed, altered, and updated; transformed maybe. The chances of this increase with the time elapsed since the original publication. The book should be read with this in mind.

Introduction
Considering Virilio's (1994) *Bunker Archeology*

An entry for the *Encyclopedia of Urban Studies* concerns 'military bunkers' (Armitage in press). It bears citation here as an introductory and contextual comment on what this book contains, and what I hope to achieve by publishing it. Created outwith the disciplines of archaeology and heritage, the entry captures the essential basis of this as an archaeological issue, albeit with clear connections elsewhere, in urban studies and sociology for example. The entry also draws in two key references which I will use in this introductory chapter, and which underpin much in the chapters that follow: one is Paul Virilio's essential and inspirational *Bunker Archaeology* (1994); the other my own more workmanlike *Combat Archaeology* (2005), which interests me here more for Mike Gane's critique in the journal *Cultural Politics* (2007), than in my own views on its significance and status.

Of military bunkers, John Armitage wrote:

> There is an increasing interest, in urban studies, sociology, and archaeology, in military bunkers. The concept informed Paul Virilio's *Bunker Archaeology* (1994), for instance, and has been significant for the Brutalist tradition of European architects, including Le Corbusier. European cultural sociology has also expanded its themes and theorizing within particular militarized landscapes and bunkered urban locations, as has contemporary British archaeology.
>
> Military bunkers are … a key component of our urban condition, if not always consciously acknowledged as such. Nevertheless, the concept has been reframed regarding the increasingly synchronized themes of postmodernity, war, and the emerging interests of the new subject of combat archaeology.

In addressing the cultural relevance of bunkers, Armitage cites my *Combat Archaeology* (op cit.) as asserting that a,

> sensitivity to military bunkers can offer an essential anchor in material culture and a stable approach to modern warfare. [One can travel] beyond recent conflict to an accelerated field of research that deals simultaneously with historical events, material remains, heritage, and human catastrophe. Such an intense combination evokes a global awareness of political events, military actions, and military bunkers. Schofield's investigation into these issues in theoretical terms and in essays on military culture and archaeological literatures, history and anthropology gracefully combine sociological discussion and concrete case studies of military bunkers as heritage management practice.

J. Schofield, *Aftermath: Readings in the Archaeology of Recent Conflict*, DOI: 10.1007/978-0-387-88521-6_1, © Springer Science + Business Media, LLC 2009

I cite this entry as it draws in important points, about the cultural values of the legacies of militarism, and sensitivities inherent within the subject matter. And these are points made, importantly, not by an archaeologist, or someone employed within the heritage sector, but a sociologist. For 'bunkers', of course, read 'all material culture of contemporary conflict', and notably all of the monuments, structures, sites and buildings that remain legible in the twenty-first century landscape, albeit as a diminishing and threatened resource (Figure 1). It is a cultural legacy that exists everywhere. Bunkers survive scattered across Albania for example, a legacy of the Cold War-era policy of defending the country, literally field by field. My acknowledgements were written in Oulu, northern Finland. As I emerged from the railway station to begin my visit, a bunker was my first sight, a vision in concrete of those classic igloos that appear in children's books and cartoons. This example is under threat, as the railway station is redesigned. Deep in the Nevada Desert are the traces of forty years of nuclear tests, the detritus (cables, objects, structures and buildings) left alone in a location where development pressures simply do not exist (Figure 2). Yet environmental degradation remains a factor here. In the suburbs of Adelaide, South Australia, I watched students excavating a Second World War air raid shelter. I could go on. Everywhere I travel there are traces, sometimes subtle and hidden away, sometimes obvious and extreme in their size and form; from the mundane to the truly and impressively monumental. It is a global heritage, and it is only at this much broader scale that the character, form and influence of militarism can be truly felt. Armitage covers this point for the urban environment. Everything he says applies equally to the wider landscape, to the 'everywhere'.

Fig. 1 Bunkers, identified here as an archaeological monument type and a sociological condition. *Photo*: author

Fig. 2 A specialist landscape of military and industrial experimentation and testing, at the Nevada Test Site (US). *Photo*: author

In addition to being an archaeology of everywhere, recent conflict archaeology is also of the everyday; the commonplace and the ordinary. Total War has ensured that we are all involved or affected, to some extent; that conflict has made some impression on the lives of each of us. Yet as heritage these ordinary places have become in some ways extraordinary, in the meanings and values they can encapsulate and convey to society. These are typically modern and supposedly familiar places that present particular and difficult choices and challenges, for heritage practitioners and the academe. There has been some questioning of archaeologies of the contemporary past, and the validity of using archaeological methods and techniques for a period so well documented, so well understood ... so familiar. This is not the place for that debate. I would simply stress what others have said before me: that the concept of a familiar past is a questionable one, and that the use of archaeology to question these taken for granteds is, to my mind, a quite legitimate and worthwhile pursuit (Buchli and Lucas 2000; Graves Brown 2001). I fail to see any distinction between fragments of prehistoric pottery and the waste flakes from stone tool manufacture, and discarded modern items or the hold-fast bolts at some military installation; graffiti and signage at military sites and medieval engravings or prehistoric rock art. We may ask different questions based on what we already know of the period, or the monument class concerned, but questions that have a basis in archaeological enquiry can always be asked, whatever the artefacts are, or irrespective of the type of site or landscape under investigation.

A fundamental difference in archaeologies of the distant and recent pasts, and what makes the latter extraordinary as an area of research, lies perhaps in what Armitage refers to as the 'sociological discussion', the variety of ways in which society accommodates militarism, and the recent history of conflict; the way society has been shaped by the two world wars and the Cold War, for example, and the way it has affected us as individuals. Most of the chapters in this book are about society. For example, I describe how in the 1950s much of the population of Las Vegas adopted the nuclear testing programme at the nearby Test Site with great enthusiasm, how some actively protested against it (Chapter 5), and how the legacy of this complex set of relations remains legible in the modern landscape (Chapter 12). The same can be said for Twyford Down (Chapter 6) and Greenham Common (Chapter 7). In both cases social benefit (as decreed by politicians) comes up against personal belief, a need to oppose authority on the basis of some deeply-felt opinion, and (often, especially in the case of Twyford Down), a close attachment to place. We can see this sociological dimension also in Sections 1 and 3 where society has made value judgements on aspects of the cultural heritage: that D-Day and Battle of Britain sites for example, as well as bomb sites and control towers, should be protected for their benefit, for their evidential value in part, but also their historic, aesthetic and communal values. These are judgements taken by government and government agencies (in these cases English Heritage) on behalf of society. It is also a view, an approach, that has been challenged, by Holtorf for example, whose viewpoint is cited in the introduction to Section 1.

Values

I want to focus briefly on what I consider the most significant of these values that are placed on sites and other pieces of material evidence: communal value. I have singled this out because, for the recent past, there is the added complication (and benefit) of having people who worked and lived in these places available to comment, to opine, and to observe and criticise the actions of those claiming to act on their behalf. In Australia this has been referred to under the heading 'social significance' or 'heritage as social action' (Byrne et al. 2001; Byrne 2008). In English Heritage's (2008) *Conservation Principles* communal value is defined as deriving from the meanings of a place 'for the people who relate to it, or for whom it figures in their collective experience or memory' (ibid.: 31). There are three categories of communal value, all of which are relevant here:

Commemorative and symbolic values: These are values which reflect the meaning of a place for those who draw part of their identity from it, or have emotional links to it. War memorials are cited as obvious examples, to which might be added the bomb sites referred to in Chapter 3, Dora (Chapter 1), and the empty ground and red earth of District Six and the District Six Museum (Chapter 2). In fact examples resonate throughout this collection. These are places that remind society of uncomfortable events, attitudes or periods of history and are often preserved precisely with the intention of doing so. The *Conservation Principles* guidance goes on: places valued for their commemorative and symbolic values, are important aspects of collective memory and identity, places of remembrance whose meanings should not be forgotten. In some cases, that meaning can only be understood through information and interpretation, whereas, in others, the character of the place itself tells most of the story (ibid.).

Social values: Social values relate to places that people perceive as a source of identity, distinctiveness, social interaction and coherence. Some such places will be modest and everyday; some reflective of regional and national identity. The Afterword is an example of the first of these, as is Strait Street (Chapter 8), Berlin (Chapter 4) and some of the places described in Chapter 13. All are modest examples of community and place. Sites surviving from the Battle of Britain (Chapter 10) and the home front in the First World War (Chapter 9) are examples of sites conveying meaning and significance in terms of regional and national identity. The English Heritage guidance (ibid.: 32) stresses that the social values of places are not always clear amongst those who share them, and may only be articulated when a place is threatened (see Read 1996). Equally these values may relate more to activities associated with a place rather than its physical fabric. This is true, for some at least, for Strait Street, though in Chapter 8 the significance of fabric is highlighted for the social values that it embodies, and (potentially at least) unlocks.

Spiritual value: Here we are dealing primarily with the spirit or sense of place, the sense of inspiration and wonder that can arise from personal contact with particular places or things. There are examples of this throughout the book as well, though the most obvious examples are in Chapter 14 where numerous instances are described of artists taking their inspiration from the militarised world of concrete installations and bunkers. This also brings us back to Armitage's recognition of the connections that exist between archaeology and the sociology of conflict, artists seeking to understand through the creative process, but also to document the impact of militarism, often in a very archaeological way by visual documentation, auditory recording and characterisation.

Documenting place can also be achieved with words, a carefully crafted passage that itself paints a picture and conveys eloquently the message and meaning of place. Michael Symmons Roberts (2001) wrote of the fence at Greenham (Figure 25):

Now, in its senility,
the base has lost whole chapters,
bailed up like a harvest
between pointless concrete posts
like standing stones.
There is no longer any difference
between outside and in.
(*The Fence*, 2001: 59).

In *The Wanderer* (ibid.: 63) he described a return to the protest camp at Blue Gate:

On a still October day – when
bonfires spin the summer into cloud –
Jason Jones, back after a decade,
takes time out at Blue Gate
on his way to Pyle Hill Woods.

Fig. 3 Exterior of the Regional Government Headquarters, Brooklands Avenue, Cambridge (UK). *Photo*: Crown Copyright NMR (AA99/01972)

Blue camp-site is black with mud
and cinders, even after all these years.
Giant concrete boulders – to ensure
no tents return – are odd now
as freak hailstones.
In the woods, Jason heaps up sticks,
tips a can of lighter fuel,
drops a match, shrugs off his olive
Air-force jacket, hangs it on the flames
as on a chair-back.

The picture is presented, the scene set, and the place enlivened.

The Cold War era Regional Government Headquarters building at Brooklands Avenue, Cambridge (Figure 3) was the inspiration behind Adrian Mitchell's (1981) poem, *On the Beach at Cambridge*, which contains the lines:

You're a poet, said the Regional Commissioner,
Go out and describe that lot.

The University Library – a little hill of brick-dust.
King's College Chapel – a dune of stone-dust.
The sea is coming closer and closer.

Again the words are narrative, describing in this case a hypothetical situation, but one that conveys the essential purpose of the building, to police and co-ordinate and supervise in a place that 'used to be East Anglia'. Again the essential character of the place is conveyed by artistic mediation.

Having established some of the conceptual, administrative and creative frameworks that underpin much of the work represented in this collection, I want now to take a brief view of the character of twentieth-century and post twentieth-century conflict, creating a more systemic framework for the examples that follow.

Settings

Returning to the influence of theorist and philosopher Paul Virilio, I will briefly examine some of the characteristics of twentieth-century warfare, notably: speed; techniques, technology and accuracy; scale; alliances; and reporting and representation, characteristics that provide a useful framework within which to further offer understanding for the examples that follow, and set recent conflict apart from that of earlier periods. Following his initial interest in warfare, expressed most cogently in *Bunker Archeology* (1994), all of these themes have been the subject of review by Virilio (see for example [with Lotringer] 1997; 2002; Gane 1999 for a review of his work and influence).

Speed

Virilio has written much on the significance of speed in understanding modern warfare (Virilio and Lotringer 1997; Virilio 2002). He has noted how the speed of decision-making is reflexive of emerging technology: in the nineteenth century battles unfolded over days, and decision making and response times could be measured in hours. The Boer War (1899-1902) is an example of the slower pace of conflict, with horses, balloons, steam traction engines, long marches, sieges and armoured trains. In the Second World War this came down to the few minutes between enemy aircraft appearing on early warning radar, and being engaged by anti-aircraft artillery and intercept aircraft. In the Cold War the significance of the three-minute warning is well documented, these warnings and practice sirens either an all-too-recent memory (although those I consulted, and who lived through the Cold War, couldn't recall whether it was three or four minutes!) or instead deeply ingrained through its use in popular culture. Reflecting this degree of instancy, Virilio and Lotringer said:

> We no longer have time for reflection. The power of speed is *that*. Democracy is that. Democracy is no longer in the hands of men, it's in the hands of computerised instruments and answering machines etc. Today there is still reaction time. It was approximately half an hour in 1961. Andropov and Reagan have no more than several minutes. (1997: 61)

Even over ten years, between the first Gulf War and the second, response times have changed significantly, from seconds to nano-seconds, as communications networks continue to improve. Virilio refers to weapons of instantaneous communication, available thanks to the development of globalised news networks and telesurveillance (2002: 49).

Speed can also be traced in the technology of weapons systems, and the ability of armoured fighting vehicles, aircraft and ships to operate within these increasingly sophisticated environments. Second World War weapons were more sophisticated than those of World War I for example; aircraft were faster, and detection systems were increasingly sophisticated as a result. Laser technology and satellites now have the capacity to deliver an immediate impact. The material record – artefacts and places – provides the physical manifestation of these developments.

Techniques, technology and accuracy

Virilio identifies three major epochs of war (2002: 6-7): the tactical and prehistorical epoch, consisting of limited violence and confrontations; the strategic epoch, historical and purely political; and the contemporary and transpolitical logistical epoch, where, 'science and industry play a determining role in the destructive power of opposing forces' (ibid.). Within this framework can be seen the development of weaponry, and its increasing significance alongside a specific 'mode of deterrence'. In the first period weapons of obstruction predominate (ditches and ramparts; armour), linked closely to the practice of siege warfare; then came weapons of destruction (lances, bows, artillery and machine-guns) which represented a war of movement; and finally 'real-time' weapons of communication (information and transport, wireless telephone, radar and satellites) that represent blitzkrieg, or total war (Table 1).

Table 1 The ages of war (after Virilio 2002)

Epoch	Characteristics	Type Of Warfare	Weaponry
Tactical and prehistorical PREHISTORY	Limited violence and confrontations	Siege	Weapons of obstruction (ditches, ramparts, armour)
Strategic MEDIEVAL	Historical and purely political	Movement	Weapons of destruction (lances, bows, artillery, machine guns)
Contemporary and Transpolitical/ logistical MODERN	Science and industry	Total War	Weapons of Communication (information, transport, telephone, radar, satellites)

These developments can be traced into the twentieth century, with many of the weapons and delivery systems now well known through media reports and popular culture. The materials used to wage trench warfare in the First World War for example are well documented, as are those of the Second World War and the Cold War. Rocket technology emerged in the Second World War through the development of the V1 and V2 unmanned weapons, used to attack British targets. After the war some of the same scientists put this experience to use in developing British and American rocket technology (eg. Cocroft 2000: 248). Blue Streak was Britain's Cold War rocket programme, given high political priority in the 1950s, and intended to be an intercontinental ballistic missile delivery system for Britain's independent nuclear deterrent (ibid.: 255-61). Sites were constructed for testing the various components, including for live firing at Woomera in South Australia. But in 1960, even before some of the facilities were completed, the Blue Streak programme was cancelled. It was thought to be vulnerable to pre-emptive strike by the Soviet Union, and there was a need to reduce defence expenditure. And that is often the way with developing technology, and with Research and Development: programmes will be realigned, intensified or cancelled dependent upon their success, the promise shown in early stages of work, on developments within science and technology more generally, and on the wider political agenda. From an archaeological point of view many of these various related programmes remain to be studied, and often without the availability of archives and oral historical evidence (but cf. Walley 2001 for an example of what oral history can contribute where those most closely involved are able to describe their experiences).

With time weapons generally become smarter, quicker and more accurate, inevitably reducing the scope for reaction time. Accuracy is important as it allows an attack to be more strategic, more focused. It can also reduce the chance of civilian casualties. The first and second Gulf Wars demonstrated how targets can be sought out precisely, and then hit with virtually no prior warning. Improvements in technology and the accuracy of weapons systems also impacts on the sophistication of decoy and deception. The use of decoys in the Second World War, in the form of dummy targets and camouflage, is now well documented (eg. Dobinson 2000a). The build-up of an allied invasion force in the UK in advance of D-Day for example made effective use of both techniques, by hiding troops and matériel in woodland close to the embarkation ports, and posting dummy aircraft and vehicles in East Anglia to draw the eyes of enemy reconnaissance. This was Operation *Fortitude*, one of the most intensively studied strategic operations of the Second World War (Dobinson 2000a: 178 ff). But even decoy and deception have changed. Now, in the twenty-first century, with weapons technology having developed beyond first-hand observation, it is also necessary to:

> Camouflage *the trajectories*, to direct the enemy's attention away from the true trajectory to lure his surveillance towards false movements, towards illusory trajectories, thanks to decoys, electronic countermeasures that 'seduce' but do not 'requite' their weaponry (Virilio 2002: 54-5; my emphasis).

Scale

Some conflict in the twentieth century has been labelled as 'total war'. War in this period typically extended beyond the confines of a discrete battle*field*, first to take in (and ultimately take out) the entire landscape (the Western Front of the First World War), extending to a global scale and incorporating sea-, air- and landscape in the Second, and impacting on everybody, however far from the front-line they may be. This developed into the risk of mutual destruction and the reality of environmental pollution (Kuletz 1998) with the physical limits extending to an infinite degree into space (with the Star Wars programme for example) during the Cold War. Again this development and increase in scale is dictated by technological capability, with the desire to win the 'space race' and take a significant lead in the Cold War being driven by military agenda, and itself driving forward Research and Development programmes.

Capability is one thing, the impact of weapons quite another, and the threat of global meltdown in the Cold War dominated many people's experience of this period. Again Virilio's progression can be seen, from hand-to-hand combat and warfare at the scale of one's own personal space, to weapons that delivered munitions from a distance and could have more impact in the sense of targeting numbers of troops and the places that contain or protect them, to those devices (now known to all as Weapons of Mass Destruction) which have the potential to be remotely triggered, and could destroy entire regions, with wider global impact. The effects of such weapons are known through testing programmes, for example in the Pacific and the Nevada Test Site in the United States, and their use at Hiroshima and Nagasaki in 1945.

Alliances

Alliances aren't entirely new, though their influence on the material culture of twentieth-century warfare has been profound. This is especially the case for the Second World War and the Cold War. It is only through appreciating the alliance in World War II for example that we can begin to understand why, three years after Britain stood alone in 1940, American troops and aircraft filled the country. Membership of NATO (the North Atlantic Treaty Organisation) as an alliance effectively against communism explains the presence of United States troops manning cruise missiles on RAF bases in England, and the presence of German 'Panzer' tanks at the Castlemartin ranges in Wales during the Cold War. The infrastructure resulting from the UK's membership of NATO was based on the operational requirements of the Alliance and not necessarily those of Britain. Furthermore, these structures were built to NATO and not necessarily British standards and specifications, points that need to be born in mind when recording and interpreting the buildings and sites that remain (Cocroft 2001; Cocroft and Thomas 2003).

All of this is true also for the Warsaw Pact. In fact some of the most interesting research questions to emerge from the Cold War concern the contrasting experiences of the period amongst those in each of these two alliances. How different were (conscripted) Russian service personnel's experiences of the Cold War from a base in East Germany (cf. Odom 1998), from that of US personnel based at Greenham Common, for example, and how might those differences be recognised through the material record?

Reporting and Representation

Developments in technology have enabled a closer proximity between events and their audience. While in the First and Second World Wars relatives and friends would watch newsreel accounts, and read the words of war correspondents in newspapers and official reports, they were some distance from the action, and from the reality of a front-line experience. But with recent wars that situation has changed. Journalists are often now 'embedded' within the armed forces, providing first hand accounts of the action. Some journalists and cameramen have died while on active service. Satellite technology enables the instantaneous communication of their reports, so action is reported in real time, into homes around the world. This proximity introduces a degree of reality to our experience of conflict, a point of particular relevance where death or atrocity is witnessed. Arguably, we are also better informed than we once were, being virtual participants now, albeit some distance removed from the action. This is true also for service personnel and their families. At the start of the century there was no contact for months and sometimes years between a serviceman away from home, and his family; now mobile telephones and welfare calls ensure close and regular contact.

What this context provides is a recognition of the types of warfare that have developed during the twentieth century, their growing impact on society and landscape, the significance of Research and Development in technological and information technological fields, and the close proximity that can now exist between the theatre of war and its wider audience (in fact with the War on Terror, the theatre now extends to all major cities, and beyond). This is a context within which to critically appraise the material culture of war during this period. It is only by understanding the significance of speed and technology for example that the significance of some of the places that drove warfare to its faster, more furious pace can be assessed, and a credible archaeological research framework established to assess that significance and research it more fully.

EMERGENCE

There is a strong historiographic element to this collection of readings, and which has influenced – to some degree – the order in which the chapters appear. It seems to be a truism that studies of any new period begin with its military structures and fabric: Iron Age hillforts and medieval castles, for example (studied at the time because they were considered military, even though current theories may present a different view), and Roman forts. It has been the same for the twentieth century, where militarism alongside industrial archaeology have led the way in developing an archaeology of (super)modernity (Penrose 2007). Of course this is a period in which the military-industrial complex ensures these two activities are hard to separate, with overlaps both frequent and obvious, not least for the Cold War, and the first industrialised conflict, of 1914-18.

But where did it all begin? Immediately prior to the outbreak of the Second World War, staff at the Ministry of Works (English Heritage's predecessors) were debating the benefit of protecting some gun emplacements from the First World War (Robertson and Schofield 2000: 18-19). The decision was taken not to, though it is significant that the debate took place at all. In the post-war years it was not until the 1960s that increased leisure time and disposable income gave rise to an ability to take interest and participate in leisure pursuits including amateur archaeology, while the passing of a generation provided the reflexive breathing space for an interest in and enthusiasm for monuments of the Second World War (eg. Wills 1985). It was shown earlier how social values often only arise when sites are threatened. It can also arise when sites are publicised in some way, and come to public attention. This happened in 1994 when D-Day sites came to the world's attention as huge television audiences watched commemorative events in Normandy, often based around the presence in beautiful coastal landscapes of ruined bunkers similar to those described by Virilio on the Atlantic Wall. This led to concern being expressed in England, and the reaction of English Heritage is described in several of the chapters included here. This may also have been the case in the 1960s, with urban expansion and the modernisation of city centres. Many Second World War sites were destroyed at this time, alongside archaeological remains of other types and categories (Darvill and Fulton 1998). Perhaps this is another reason why, at this time, the interest in Second World War remains in Britain surged.

Archaeology and heritage management practice had already become a far more inclusive and broader church by the time the Cold War ended, an ending that was reflected in government policies towards closure and contraction amongst the armed forces, both in Britain and overseas. The existence of a threatened resource and the willingness to create a response to that threat were uppermost on heritage agenda. In the spirit of informed conservation, Cold War sites were urgently assessed and evaluated to the extent that informed judgements could be made on future priorities, for research and conservation. Often decisions come about by accident rather than design, but at least understanding provides weight when decisions and answers are required. Archaeology and heritage practices now comfortably deal with contested pasts, the modern era and sites with particular troubled histories. A broader archaeology, and heritage management practices that have become inclusive not exclusive, that are starting to address the everyday and the ordinary in addition to those 'critical assets' only of 'national importance', has meant that many of the issues addressed by the chapters in this collection are no longer as radical or leftfield as they once might have been. The extraordinary has become ordinary over the past decade. This book charts that process, at least so far as I saw it.

Scope and Direction

As I hope will be obvious from the preliminaries, this book is not intended as a 'read-through', but rather a collection of scripts and images that can form the basis of learning, and of research, for student coursework and projects for example, whether in military archaeology, heritage studies or the historiography of archaeology and heritage over the past 10-15 years. Some may also read the collection as a matter of general interest, for example charting the continued development of conflict archaeology in the years c.1995-2007, years of change, for the better I would argue.

All of the chapters in this book, excepting the section introductions, the Afterword and all other preliminaries, have been published previously, often though in rather obscure and inaccessible places. Here what was always envisaged as a collection of related works is finally brought together, and contextualised, given new meaning through the sum of its related parts. All of the chapters have been revised slightly from their original form: some corrections have been made, one or two updates included, and about half of the original illustrations replaced or removed. But essentially the chapters are retained in their original form. As I have explained, the significance of the collection is that it charts a progression which encapsulates a period in which the archaeology of recent conflict became accepted, mainstream and professionalized. As I said in the Preface, this was not my doing; I was merely a player, along with many others, most of whose names and key texts appear in the References and throughout the text. That said, I have been closely involved since 1995, and the chapters included here represent projects, events and the emergence of ideas and priorities from that point on. The collection inevitably represents a personal perspective, coincident with a corporate view in some cases, but one that sits within a broader narrative of archaeological and heritage management practices during the late twentieth and early twenty-first centuries.

NOTE: A version of the section 'Settings' first appeared in the opening chapter of Schofield J. 2005, *Combat Archaeology: Material Culture and Modern Conflict*. London: Duckworth.

Section 1
Frameworks in Conflict Archaeology

This first section provides context to what follows, a definition of frameworks grounded in the principles and practices of cultural heritage management, and of historical and contemporary archaeology, at least as these exist in the UK where the majority of my work has focused. The section is a scene-setter therefore, a prologue, a discussion of general issues that applied at the time of the chapters' original publication just as they do today. Ultimately though, the three chapters included here owe most to the notion of informed conservation, by which is simply meant understanding a subject sufficiently to make informed decisions about it – in heritage as in life! Thus, the essays encapsulate the wider principles of applied research: research undertaken for a particular outcome, and often funded only for that reason. In the past, essays and books of mine have been criticised for being too closely driven by heritage agenda (as if heritage and matters for the academe are somehow separate – which I would dispute). *Combat Archaeology* (2005), for example, was described in the magazine *Current Archaeology* (Issue 201, 2006: 494–5) as 'a rather dry manual for heritage managers' and being infused with a 'bureaucratic spirit'. If that is true then I make no apology for it. Given where I work and what I do, it would be difficult to conduct or write about this research in any other way. The bureaucratic spirit is evident here too, though its influence diminishes as the chapters pass.

All the three chapters in this section are about management – about why conflict archaeology matters, for whom, and what can be done about it. All the three were published in edited collections arising from the heritage-based conferences and seminars in which the original presentations were made. Therefore, they have that in common also. Yet they were written at very different times, and for different reasons. Chapters 1 and 2 were included in a collaborative seminar series involving English Heritage's Education Service and the publisher Routledge. 'Conserving Recent Military Remains' came first, being a late addition to a seminar held at the Society of Antiquaries of London in October 1997 on the subject: Presentation and Preservation: Conflict or Collaboration? (eventually brought together under the published title: 'Managing Historic Sites and Buildings: Reconciling Presentation and Preservation'). 'Jessie's Cats and Other Stories' was originally presented at a Heritage Interpretation seminar, also held at the Society of Antiquaries in 2000.

As I said in the original publication of this chapter (2006: 142), Jessie's Cats follows on directly from the earlier chapter, taking some of the broader issues and dealing with them in a more personal, a more intimate way. The third chapter, 'Monuments and the Memories of War' came between the two, in 1999, presented as a paper at the fourth World Archaeological Congress in Cape Town. Indeed it was at this conference, and the related visits to District Six and Robben Island, that inspiration came to research and write 'Jessie's Cats'.

These short section introductions are intended as simple commentaries on the essays themselves, giving full details of the original publications, and the research context from which they originate. But they are also an opportunity to comment briefly on how these essays were described at the time, by the editors of collected works for example, or in a few cases, by reviewers. I do read and consider carefully the reviews of books and essays I have written. But I do not actively search for them – if they come to my notice I read them; if not, I live in ignorance of their existence. My section introductions should be read with this in mind: that the critique is, of necessity, a selective one, and potentially skewed.

The three essays in this section each concern what we do with recent conflict archaeology, or matériel culture (Schofield et al. 2002), or 'combat archaeology' as I have also called it (Schofield 2005a), but also how we do it and why. The first essay is more about whether we should, and why such concrete ruins matter to society at all. The second concerns how we interpret these troubled or dissonant sites and stories; and the third is specifically about place and memory, but in a general sense, preceding the more detailed studies that follow in Sect. 2. Together these essays give an impression of how a state heritage agency was thinking at around the time of the millennium, views that can be compared to the post-disciplinary, arguably more broad-minded and post-disciplinary optimism of later years, represented for example in the two essays comprising the final section of the book. Indeed the dates of publication for all the chapters in this collection are directly relevant to their reading, analysis and (hopefully) their critique and decon-struction by those that read them.

But let us begin with Chap. 1, and the first essay in this collection to be written. Introducing the various chapters in 'Managing Historic Sites and Buildings', in which Chap. 1 appeared, David Baker described my chapter as providing,

> a fitting final contribution to this book and the immediately pre-millennial year in which it is published. The twentieth century is already self-consciously history, and one of its defining characteristics has been human conflict and bloodshed on an unprecedented scale, with harnessed technology magnifying the mayhem and making everyone more aware of it. Despite not having been invaded in recent centuries, Britain is not short of its military remains, and those from the last great conflict are now moving steadily into that hinterland of recollection by a shrinking minority of the population, many with painful personal memories, while an inevitably increasing majority associated them primarily with what they see on television or read in books. War is the most destructive agent of change in the historical process: those of its relics not destroyed in acts of disassociation soon after the event are easily pressed into the service of a partisan view about a conflict or all warfare itself. (1999: 18)

Baker also highlights the treatment of monuments and the process of change. He comments, notably, how in one sense,

> Preservation as 'monumentalisation' inevitably detaches the survival from its social and economic context, which then continues to evolve; in another sense, however, the monument has a new role in that context as a relic that helps define the present, and how it views the past, (1999:9)

a point, he suggests, is starkly and poignantly illustrated by the case of recent military remains.

District Six is the first and major case study in Chap. 2, and it was the inspiration for the chapter as a whole. But it was written, inevitably and unavoidably, from the perspective of a visitor, a traveller unfamiliar with the intricacies of the political situation in South Africa under the Apartheid regime (recognising though that my partner is from South Africa, and was involved in anti-Apartheid protests and rallies at the University of Cape Town and in the city in the later 1980s). There will certainly be a naivity behind my interpretation (I would hesitate to call it insight) and for other more informed views one should read works by Martin Hall, for example, some of which appear in the list of references. The chapter is not only about District Six, however. It uses that and other examples to emphasise my main point, being the importance of personalisation in presenting and interpreting conflict heritage sites if the greatest impression is to be made on visitors. It is also to large extent what gives the sites their significance: '[These] are clearly not places of outstanding natural beauty or human artistry, although ingenuity, ruthlessness and inhumanity may all be present, along with political ideologies and particular views of the world' (Hems 2006: 6). Furthermore the chapter emphasises, as the editor of the volume Alison Hems points out, that at this point interpretation becomes an exciting, if not dangerous activity (ibid.). She goes on:

> Much of this [materiality] will be invisible without interpretation, although the building may retain its own power, and the resonance of memory. For those that lived through the events that took place there, the preservation of the building as a memorial may well be enough. Yet recent events pass rapidly into history, and the carefully preserved structures and original features can equally rapidly lose their significance – unless one knows what it was, and can acknowledge changes in significance and understanding as time passes. ... In this way [the symbolic value of Orford Ness, a Cold War experimental site in Suffolk] will be maintained for future generations. Yet the symbolic value of Cold War sites is already different for different audiences, and will change again as the rival perspectives of, say, nuclear protestors and government scientists pass beyond living memory and into history. A policy of non-intervention goes a long way in preserving 'an aura of mystery' or in presenting cold, grey structures [from] which visitors depart feeling 'chilled', but it may not offer them the insights they need to place the monument in its broader context or to deal with their emotional response to its creation and survival. (Hems 2005: 6, my brackets).

'Jessie's Cats' was a contribution to a volume on heritage interpretation, and the only contribution in that collection that dealt directly and explicitly with the troubled past. Yet my chapter ended there, as it does here, with recommendations or at least ideas that might contribute to good practice, not only for conflict heritage sites, but the historic environment generally. Hems made the same point, noting how,

effective interpretation must involve audiences in hearing and telling past stories; it emphasises human experience, and places it at the core of those stories. Adopting such approaches,
in the context of landscape management or the conservation of the built heritage is notori-
ously difficult to do (ibid.).

The third chapter, 'Monuments and the Memories of War' is different, in that it
was included within a session which focused entirely on this subject, arguably for
the first time at a major international congress, and in a place which had seen its
fair share of troubles. The session was co-organised with two north American
scholars, and none of us had been to South Africa before. Indeed the majority of
our contributors had never been there either. The session was one thing. It was well
attended and regarded sufficiently by others to merit publication in the One World
Archaeology Series (Schofield et al. 2002). But the wider experience was very different, not least due to our collective experiences of post-Congress tours to Robben
Island, District Six and one, arranged specifically for speakers in our session, of
Cape Town's townships. For many it was our first personal experience of poverty on
this scale, and of the history of conflict that has created the present situation.
'Monuments and the Memories of War' should be read with this wider experience
of the conference in mind.

Given that much of the emphasis of this third chapter concerns the value of
monuments and their significance for remembering, or for not forgetting, Cornelius
Holtorf's review of *Matériel Culture* (2004a) is worth citing here. He makes some
important points, here and elsewhere, on the benefits of and concerns about
preservation. On reflection I am sure Holtorf is correct in stating that in this,

No one (meaning amongst the contributors to the book) seems to be concerned with the
specific social constitution and performances it takes, or will take, to appreciate archaeological sites of twentieth-century conflict in a meaningful way. Many of the authors, as well
as the editors, seem to take for granted that such sites, once preserved, will function as
historic mnemonics of some kind, for the only reason that they were once meaningful in
the past. But will future generations at these places really remember twentieth-century
conflicts? Or might they instead remember the twentieth-century political disputes over
their heritage status, and thus, at most, remember remembering twentieth-century conflict
(2004a: 317).

He goes on:

Less (preservation) can be more (memory). Moreover preservationists are running the
risk of reproducing the logic that governed many human rights abuses, wars and genocides
in the past. Should heritage too be about controlling material resources, claiming disputed
places, and manipulating collective memories? Is this the final *Materialschlacht* and what
is the world going to look like when the battle is over? (ibid. 318).

These three essays are reproduced by kind permission of the original publishers.
Chapter 1 was originally published as Schofield J. (1999): Conserving recent military
remains: choices and challenges for the twenty-first century, in Gill Chitty and David
Baker (editors), *Managing Historic Sites and Buildings: Reconciling Presentation
and Preservation*, pp. 173–186. London: English Heritage and Routledge. Chapter
2 appeared first as Schofield J. (2006): Jessie's Cats and other stories: presenting and

interpreting recent troubles, in Alison Hems and Marion Blockley (editors), *Heritage Interpretation*, pp. 141–162. London: English Heritage and Routledge. Chapter 3 first appeared as Schofield, J. (2002): Monuments and the memories of war: motivations for preserving military sites in England, in John Schofield, William Gray Johnson and Colleen Beck (editors), *Matériel Culture: The Archaeology of Twentieth Century Conflict*, pp. 143–158. London and New York: Routledge. One World Archaeology 44. The chapters reproduced here differ somewhat from the originals. The illustrations are only occasionally the same as those of the original.

Chapter 1
Conserving Recent Military Remains: Choices and Challenges for the Twenty-First Century

The mountain Kohnstein is anhydrite but Dora is quicksand. It sucked its slaves into the earth; it sucked its Nazi guards into the abyss of inhumanity; it sucked its scientists into the blindness of goal without consequence, accomplishment without accountability.

It has continued to absorb all who come to the tunnels. It has trapped the researchers who have come to study; it has trapped the historians who have come to write; it has drawn the visitor who has come to wonder at the wonders and the cruelties that have taken place in this strange and hallowed ground.

Anhydrite can be mined, as history can be mined; the tunnels of Dora go into the earth forever (Gilens 1995: 113).

Recent military remains are, by definition, a new dimension to the heritage, one that provides a significant and challenging addition to our historic cultural resources. The cultural value of these remains, and the nature of the challenge they present to those charged with their conservation, protection, presentation and marketing, form the subject of this opening chapter. Dora is indicative of the chapter's central theme: that recent military sites often evoke a depth of feeling rarely seen on other types of site (excepting perhaps the scenes of industrial disasters). That emotional charge is expressed here by one of the first post-war visitors to Dora's underground world, where slaves from Buchenwald were brought to work on Germany's V2 rocket. Dora also highlights the conservation dilemma: the tunnels were sealed at the end of hostilities and they survive as left, as a monument and a memorial, a testimonial and a shrine (Cocroft 2008: 27–31).

Although recent military remains have been of interest to enthusiasts and amateur archaeologists over at least 40 years, a professional concern and popular support for the physical remains of twentieth-century conflict has developed only more recently. The reasons are becoming increasingly clear, and extend beyond the mere fact that 'heritage now spreads into yesterday' (Lowenthal 1996: 17): the nostalgia surrounding the 50th anniversaries of VE and VJ days; the reflective mood that accompanied the millennium; the development pressure placed on military remains, for instance through the Ministry of Defence's disposal programme, made necessary by downsizing the armed forces, and operational changes brought on by technological advances in warfare; and changing perceptions of what constitutes the historic environment, and the view that it should be regarded holistically, a key

J. Schofield, *Aftermath: Readings in the Archaeology of Recent Conflict*,
DOI: 10.1007/978-0-387-88521-6_2, © Springer Science+Business Media, LLC 2009

principle in the notion of sustainability (English Heritage 2008b). There are certain considerations, however, some unique to recent military remains, that require that these sites command special treatment, and it is these considerations which form the basis of this chapter. First is the philosophy of preservation: can we view objectively that which is so recent; and should we be preserving sites which can evoke such painful memories? Second, how should military remains and other structures synonymous with the two World Wars and the Cold War, as well as those of civil conflict, be presented to a multi-cultural and multi-national audience embracing both veterans and the very young? Finally, if we should preserve some of these sites and structures, on what basis can such a selection be made, and what form of protection is most appropriate for conserving, on the one hand, redundant military structures, and on the other, the many buildings which remain in use?

Hallowed Ground: Principles of Conservation

The recent past is today considered as much a part of our heritage as are more distant periods (e.g. Hicks and Beaudry 2006), and as we move into the twenty-first century, its status as history and its cultural significance will become even more obvious and more consensual. This concerns not just military remains and the traces of industry, but the entire landscape (Penrose 2007); as Samuel has said (1994), the notion of heritage is serving to modernise and update what constitutes the historical, as well as extending its social base. English Heritage views recent military remains as an important part of this wider heritage. Such places exist as touchstones or markers of global conflict that will surely come to characterise the twentieth century, described variously as an 'age of extremes', and as the most terrible in western history. These were after all momentous events which shaped nations; they made the modern world.

Various arguments support the preservation of recent military remains. There is a view that selected remains of the two World Wars and the Cold War must be preserved, in order that we 'retain our sense of history', as well as giving character to our towns and countryside – the sense of place and community which held such significance during the war years. Furthermore, these remains play a significant role in British history – in some parts of Britain, the changing character of defence systems, from the medieval period to the middle of the twentieth century, can be viewed and readily appreciated within their physical and strategic context (but cf. Stocker 1992). Also, and importantly, military remains – combined ideally with the testimony of those involved – give archaeologists the opportunity to 'turn the dead silence into an eloquent statement of experience' (Carman 1997: 2). In this regard, military remains are also significant in education. The remains are everywhere – in town and country – and it would seem that, increasingly, their educational and recreational value is being recognised (e.g. Planel 1995). Then there is the economic value. Military sites and museums, local and national, and in particular the Imperial War Museums in London, Manchester and Duxford, attract considerable numbers of visitors including many veterans who return to their wartime bases, of which Duxford was one. Finally, there is an emotional value, connected with bereavement, remembering and commemoration (Tarlow 1997), though arguably this has greater relevance at the scenes of conflict and atrocity (concentration camps such as Auschwitz – inscribed a World Heritage Site – and the First World War battlefields for instance) than in areas beyond the war zone.

There are alternative views to the suggestion that such sites should be retained or preserved. The first concerns sustainability (English Heritage 2008b), and specifically the role of the 'community' in defining its own 'critical assets'. Put simply, sustainability requires some mechanism, independent of the heritage agencies themselves, by which to determine whether items of heritage capital are sustainable (Lowenthal 1996: 21). Where they have neither community support nor passive acceptance, this is unlikely to be the case. There is certainly a view that support for recent military remains is to be found only among special interest groups, though there is growing evidence (referred to later) that appears to contradict this. A second view is that monuments of war should be removed as unsightly and unstable reminders of a sad and violent past. Although unusual nowadays, this view perhaps has some link to the idea that, while continuing to honour the war dead, the millennium may be a suitable time to consider modernising the act of remembrance, perhaps even reducing the central significance of war memorials and the symbol of remembrance – the poppy – in favour of something more forward-looking and 'less triumphal' (McCrum 1997). The largely favourable reaction of veterans to the recently reformed Spice Girls 'promoting' Remembrance Day in 1997, wider participation in the two-min silence, and the seemingly greater enthusiasm of the very young, who 'want to know', and who actively participate in the collective act of remembering the war dead, may indicate some support for this view.

Three examples illustrate some of the difficulties military remains present in the years immediately following conflict, occupation or repression, and the first example, the Berlin Wall, exemplifies this as well as highlighting a dilemma between conservation and consumption (Fig. 4).

The Wall was built in 1961 in the stalemate phase of the Cold War, to stop mass emigration from east to west, and in the propaganda of the German Democratic Republic, as a contribution to World Peace: 'a foundation stone for the success of our policy of relaxation and peaceful co-operation' and as an 'anti-fascist protection wall' (Baker 1993). Prior to its removal in 1989, the Wall was undeniably *the* symbol of the Cold War, its significance and power felt the world over, as well as being the defining feature of Berlin (Schmidt 2005). Within seconds of the first hammer blow, television showed the world the Berliner's instant reaction – to take physical possession of it, climbing on it, then hacking at it. Watching events unfold, we were experiencing a defining moment of the twentieth century. It is easy to understand the reaction to tear it down: it separated many families. Nevertheless, as it came down, nostalgia combined with a conservation ethic in reaction to the overwhelming consumption of Berliners.

Alfred Kernd'l, Berlin's chief archaeologist, said this in support of preservation: 'It is typical for us Germans that at the end of an historical era we want to rip

Fig. 4 The Berlin Wall, looking east (1983): symbol of the Cold War and the subject of a significant conservation dilemma to those charged with its management. *Photo*: author

everything down and forget it ever happened. It occurred with the Nazi sites, now it's happening with the Berlin Wall' (Baker 1993: 726).

Arguments for preserving parts of the Wall were also offered by the Green Party on the basis that it was essential for understanding a critical 30-year period in the city's history, as well as being symbolic of one of the most important periods in world history.

Eventually it was proposed that three main sections of the wall be retained, one of which was controversial as it ran alongside the former Gestapo and SS HQ of Himmler and Heydrich – at one time the most feared address in Berlin – this begging comparison between the horrors of the Nazis and the Stasi. One view was that visitors should be confronted by the inter-relationships of German history; another was that preserving a section of wall adjacent to the Gestapo HQ would serve only to relativise and dilute the crimes of the Nazis (Kernd'l, quoted in Baker 1993: 727). It is tempting to suggest that, in future, the interest in this monument will be as much for its demolition by the people, as for its limited physical presence. Indeed, the heritage management regime could almost be seen as representing bogus interference in the historical process – an agency acting on behalf of reactionary forces, and against the community will (David Stocker: personal communication).

A further political dimension here is that preservation – especially of Nazi landmarks – can lead to their becoming the focus of modern extremist – in this case neo-Nazi – organisations. The Berghof – Hitler's alpine retreat – is an example: demolished by the Bavarian authorities some time after 1945, it nevertheless became a place of pilgrimage. The present and controversial plans to construct a documentation centre on the site incorporating what remains of the underground bunkers, have the intention of retaining its historic significance while having a use that discourages the neo-Nazi presence (Traynor 1997). One part of the complex – the Eagle's Nest – where Hitler received Chamberlain and Stalin, is now a mountain-top cafe.

A second example is the Channel Islands, under German occupation from 1940–45: Hitler intended to make the islands a new Gibraltar, German for all time, thus committing vast quantities of steel, concrete and labour to their fortification. Immediately after liberation, there were stringent efforts to eliminate or hide these reminders of Nazi occupation: much was removed by the British liberating forces; what was not transportable was sold for scrap. But once the more moveable features had been dealt with, valuable items sold, and memories began to fade, the pace slowed almost to a standstill. On Guernsey, many of the more robust fortifications, such as the observation towers, survived, and 14 of the most significant sites, determined from an island-wide survey, are now protected under their Ancient Monuments and Protected Buildings Law (Fig. 5).

These remains now have economic benefits – some sites are marketed by the Tourist Board under the umbrella 'Fortress Guernsey' – and their physical presence (and indeed that of re-enactment groups in Nazi uniforms) seems not to bother the islanders of today, irrespective of their generation. One observation tower and a

Fig. 5 Naval observation tower at Pleinmont: one of 14 German fortifications (or groups of fortifications) on Guernsey (Channel Islands) protected under Ancient Monuments and Protected Buildings legislation. *Photo*: author

coast battery now provide holiday accommodation, the latter on the nearby island of Alderney.

One might compare this situation with that of Vietnam where western visitors are advised not to wear shorts, not because of tropical insects or risk of infection, but because of history – shorts remind the Vietnamese of the French, who wore them during the period of their occupation over 40 years ago. But note also the contradiction (albeit based on economic expediency): Webster (1997: 164) has reported that, in the Vietnamese language, the word for 'American' is the same as the word 'Beautiful'. As a Vietnamese put it: 'We are a practical people, and we remember only what we can use of the past. Now we think the Americans can help us. So... we love the Americans.'

Finally, Denmark: Here a 'special emotional problem' is described concerning the works of World War II.

> These are reminders not only of a military occupation but also are symbols of the brutality of the Third Reich as experienced by the Danish people. One can, naturally, choose to ignore them or remove them from sight, but an alternative is to let these works remind us, and coming generations, what Nazism and the Third Reich stood for (Ministry of Environment 1994: 41).

In all, the Germans built more than 6,000 bunkers for Denmark's coast defence, yet very few of those are visible today. As on Guernsey, those which are visible are popular as visitor attractions, and are used in education. Many of the remainder were stripped of equipment after the war and covered over; some survive buried beneath sand dunes (Ministry of Environment 1994: 39).

There appears therefore a sequence of attitudes and motivations following occupation and the cessation of hostilities, and it is interesting to note in this context of time-lapse a clearance initiative in Kuwait seven years after the end of the First Gulf War: here, by 1997, 112,959 bunkers had been destroyed, 213 miles of trenches filled in and 243 miles of earthwork berms levelled by the company employed to clear the war zone (Webster 1997: 229). We hear much about (and applaud) the clearance of land mines, but the clearance of fortifications has a different motivation altogether, driven more by psychological than health and safety concerns. By comparison it appears that clearance of the Gulf's desert landscape was much more systematic than was the case with Guernsey in the years immediately following World War II, and much more permanent than the clearance of German sites in Denmark: indeed there seems a real possibility that all physical trace of the First and Second Gulf Wars will be gone within a short time. Is there not a case for preserving a selection of these structures as monuments to the latest conflict in the history of this troubled region, or is the materiality of media and new technology a sufficient record to remind future generations of this particular episode? As with the Berlin Wall, any attempt to secure preservation may not have 'community' support, mainly given the recent occurrence of events. But it can be argued that, despite the charge of interfering with the historical process, we (the heritage sector) have a duty to preserve a selection of sites for the benefit of future communities in that region and more widely.

'Clear-Cut, Successful in All Respects': Presenting Conflict as Heritage

As we have seen, there is a duty on those charged with presenting recent military sites to balance numerous responsibilities: to remember the fallen; to avoid trivialising contributions to the war effort; but also (I would argue) to ensure some emotional engagement with the subject. David Uzzell (e.g. 1989) termed this 'hot interpretation'.

The controversy that often accompanies interpretation is exemplified by events surrounding the proposed exhibit at the Smithsonian (Washington DC), to commemorate the bombing of Hiroshima – described by the weaponeer who released the bomb as 'clear-cut, successful in all respects'. In his book, Harwit (1996) chronicles, with a combination of dispassion and anger, the long evolution of the museum's plans for an exhibition in observation of the fiftieth anniversary of the bombing, for which the centrepiece was to be the Enola Gay, the B-29 aircraft that dropped the bomb. Over a ten-year period, the aircraft had been painstakingly restored and well satisfied the criterion of historic importance necessary for its inclusion in the museum's Air and Space Collection. However, unlike the other hardware on display, the Enola Gay was not there as a triumphant manifestation of higher, faster or further, but rather because it initiated the age of nuclear weapons, killed some 100,000 people and hastened the end of the Second World War. So, not surprisingly, controversy surrounded the question of what the museum's visitors were to be told about this aircraft and its place in history. On one side of the debate were veterans who felt the plane should be displayed 'proudly'; they demanded an approving if not celebratory observation of the plane's wartime feat. Harwit, however, wanted to infuse awareness that the bomb had caused damage and suffering. The veterans were incensed by plans to display charred artefacts from the bombing: a child's lunch box; a clock that stopped at the moment of detonation. They and other critics said the proposed display 'sentimentally ignored Japanese culpability and cruelty in the war'. In the end the exhibit was cancelled, leaving only the forward section of the fuselage on display with scant reference to its historic role. The balance between critical analysis and honouring and commemorating valour and service, it would seem, is a fine one.

Conservation Practice in England

Turning to more practical matters, and specifically the question of how recent military remains can be conserved, I will outline the staged approach taken by English Heritage towards the evaluation of recent military sites in England in the period c. 1996–2005, something I will also return to in later chapters. There are two main stages to this: First, how to assess the resource – how we can begin to appreciate its extent and diversity, and from that the relative importance of its component parts.

Second, what options are there in England for its future management? Again, some examples will follow general discussion.

In England, as elsewhere, conservation practice is based on the principles of sustainability and informed conservation: specifically, that not everything can (indeed should) be preserved in situ, and that decisions on future management must be based on the best possible information. Some elements of the historic environment, it is argued, should be preserved at all costs (these are our so-called critical assets, deemed to be of great value and irreplaceable); some will be subject to limited change; while some can be exchanged for other benefits. In dealing with recent military sites, English Heritage's understanding was improved by commissioning a national review of the subject (cf. Dobinson et al. 1997 for details). By consulting primary sources held at the National Archives (formerly the Public Records Office), work has been undertaken and thematic reports produced for most major classes of recent military site (Table 2). Each report contains an account of the historical context relevant to the site type, details of typology, chronology, as well as (with two exceptions, see Table 2) gazetteers detailing how many sites there were, where they were (usually to the accuracy of a six-figure grid reference), when they were there and what they looked like (often with ground plans or photographs). For two types of site (anti-invasion measures of the Second World War and civil defence), sites were too numerous to be accurately recorded through primary sources. For all of the site types listed in Table 2, our understanding has been transformed by this survey: primary sources do give a virtually complete account of sites as built, and aerial photographs were then used to assess the likelihood of their present survival. Finally, for the Cold War period, the survey of primary sources was restricted by the Thirty Year Rule to the period up to 1968, but here work by English Heritage's survey teams proved invaluable, recording structures which exemplify the main site types, including missile launch sites, radar and communication installations, military bomb shelters, peace camps, research and manufacturing sites and the

Table 2 Scope of the survey of primary sources undertaken by English Heritage

Site types	Dates	Distributional information
Anti-aircraft artillery	1914–46	complete
Anti-invasion defences	1939–45	representative
Bombing decoys	1939–45	complete
Operation *Diver* sites	1944–45	complete
Operation *Overlord*		
Preparatory sites	1942–45	complete
Coast artillery	1900–56	complete
Civil defence	1939–45	representative
Radar (inc. acoustic detection)	1920–45	complete
Airfields (inc. airfield	1914–45	
defences)		complete
Cold War sites	1947–68	complete

'Little America' architecture of the large US Air Force bases (Cocroft and Thomas 2004; Cocroft 2001). Primary sources are still relevant here and have a clear and significant role in understanding and reinterpreting Cold War history at a global level (Gaddis 1997).

In conserving historic fabric, English Heritage has traditionally made the distinction between structures and sites that are best served by their future as monuments, and those for which use, or adaptation for beneficial reuse, is appropriate. In military terms, this distinction broadly corresponds to that between the so-called teeth and tail of the armed forces. Buildings of the support services – the tail – are often generalised structures and as such will often continue in some form of use, while the teeth – including fortifications – are now monuments for which beneficial reuse is generally difficult to envisage, given their design as specialised and functional structures. Presently, at the time of writing, this system (which also includes separate provision for conservation areas and battlefields and landscape registers, for example) is changing, with a single designation and a single, more straightforward and more transparent and democratic system replacing it.

The development of the English scheduled monument and listed building systems, as they will continue to be for only a short while longer, occurred through the late nineteenth and early to mid-twentieth centuries and was based around a partnership between these complimentary approaches, the distinction being made between: (1) the management needs of those critical assets whose preservation takes precedence over their use, and which may have no use except as monuments (these are presently covered by scheduling under the terms of the 1979 *Ancient Monuments and Archaeological Areas Act*), and (2) those buildings whose conservation value is best safeguarded by retaining them in use, whose sympathetic use is, in fact, a form of conservation (for which listing is appropriate, under the *1990 Planning [Listed Buildings and Conservation Areas] Act*). This distinction has resulted in two separate sets of management controls, tailored to meet the needs of their respective constituencies. Scheduled Monument Consent, for example, (the controls which ensure the care of scheduled monuments) makes the assumption that efforts will be made to preserve the designated structure in situ and in more-or-less the state in which it came down to us. Its aim is preservationist. On the other hand, listed building controls are more flexible and based on the assumption that some measure of alteration and adaptation may be necessary in order that the building can maintain a viable use and will retain its value as a capital asset.

Examples illustrate the extremes. The gas warfare testing trenches on Porton Down (Wiltshire) are a Scheduled Monument. With an overall diameter of nearly 400 m, these concentric trenches were dug in 1916 during experiments with gas and other forms of chemical warfare. Their concentric design enables gas to be released from the inner trench and its effects tested on personnel in the outer, irrespective of wind direction. The trenches survive as earthworks. The monument is demonstrably of national importance (it is likely to be unique in Britain), and is clearly one for which the strict controls of Scheduled Monuments legislation were appropriate. Military buildings which are best managed through continued use, on the other hand, include many of the UK's best-known architectural landmarks, such as Horseguards in Westminster, and the Royal Naval College at Greenwich. Clearly such buildings should remain in use, and for that listing was the appropriate designation. This is also the case for most aircraft and airship hangers. At Calshot, for instance (Fig. 6), the First World War flying boat hangers are listed, and now contain one of Hampshire County Council's outdoor leisure centres, including a dry ski-slope and cycle track. The Henrician fort of Calshot Castle, managed and presented as a heritage attraction by English Heritage, is a Scheduled Monument.

Under this regime, therefore, the various designations could work together, a point most clearly illustrated by English Heritage's work on airfields and airfield structures, where – through close co-operation with Paul Francis and the Airfield Research Group – information has been gathered on relative significance, both of the airfields themselves and the structures and sites within them. Designation on such extensive sites, many of which remain in use, can be a complicated affair: in a case like Bicester (Oxfordshire), listing was appropriate for the majority of buildings which remain in use on the Domestic and Technical sites; scheduling was applied only to the well-preserved defence structures which surround the airfield (including blast shelters, air-raid shelters, pillboxes); and conservation area status covers the whole site, with the intention of retaining the overall character of arguably the most complete of the RAF's 1920s' bomber stations, one of only three in England to have retained its grass

Fig. 6 Calshot Castle (*centre*) and the flying boat hangars (to its *left* and *right*), now managed through scheduling and listing respectively. *Photo*: author

field, and the most complete airfield site in Britain to predate the 1930s. Conservation area status has been used before in this context, at Hullavington (Wiltshire) for instance, and Biggin Hill (Kent). Clearly, a unified list with a single form of designation and clearer procedures will be beneficial in cases such as these.

Also relevant is the implementation of *Planning Policy Guidance Note 15* (PPG-15) and PPG-16 in conservation practice (documents and practices that are also currently under review). PPG-15, published in 1994, is important as it provides a full statement of Government policies for the identification and protection of historic buildings, conservation areas and other parts of the historic environment. It also explains the role of the planning system in their protection and complements the guidance on archaeology and planning given in PPG-16 (published in 1990). PPG-16 has had the effect of greatly increasing the ability of the planning process to protect and manage archaeological sites, and in these terms the current work on recent military remains is beginning to assist the implementation of PPG-16 at a local level: in North Yorkshire, for example, as a condition of planning permission, developers are now required to record military sites. An example of this has been the thorough photographic recording of two RAF camps associated with Scorton Airfield in advance of redevelopment.

Conclusion

Recent military remains are an integral part of the historic environment, and one which appears to have growing public support. The various branches of the Imperial War Museum attract large numbers of visitors, as do many of the locally managed attractions around the country. In England, as elsewhere in the UK, The Defence of Britain Project proved successful in harnessing that enthusiasm and making good use of it in its national recording programme. Schoolchildren seem genuinely interested in the subject and want to know more. The media have covered many items on local and national initiatives. Until recently, our understanding of the material remains in England representing the two World Wars and the Cold War has been poor. However, one of the principles of work being undertaken by English Heritage is that any statutory designations must have a credible basis. Above all else, the research outlined here aims to provide that.

Within the profession there seems little doubt these days that a selection of these monuments, teeth and tail, and reflecting the changing nature of conflict during the course of the twentieth century, should be preserved for the future, to serve as touchstones for what is already being described as a calamitous century, an age of extremes. With wider support, English Heritage is moving towards this position. Other countries are doing the same. But it is not an easy subject, either in terms of conservation practice or philosophy. With war and conflict, the choices and challenges are greater than ever, and the pressure on those charged with its preservation and presentation to 'get it right' is prodigious.

Chapter 2
Jessie's Cats and Other Stories: Presenting and Interpreting Recent Troubles

Every Friday after work Jessie would collect a standing order of minced meat from her butcher in Hanover Street. She bought the mince specially to feed the cats in our area. Come rain or sunshine, Jessie turned up every Friday afternoon in the lane behind our house. Here she would stand on tiptoe to reach onto the high wall where all the neighbourhood's cats were gathered.

The cats loved her. They turned up in different shapes and sizes, colours and temperaments. Some of them were rough and ugly, but when Jessie fed them, they all behaved like sweet, adorable kittens… One Friday afternoon, Jessie, the 'fairy godmother of the cats', failed to arrive. The hours passed and the cats waited and waited. They all lingered, clearly hoping that she was merely delayed.

But Jessie never turned up to feed them…

[…]

Later (after the forced removals under the *Group Areas Act, 1966*), rumours started going around that Jessie had (returned to the area and) been seen feeding the cats. Some of the people who were still living in District Six swore that one could on a Friday afternoon sometimes catch a glimpse of a silhouette standing up on tiptoe to feed the cats on the wall.

Many Friday afternoons around five we looked out for the silhouette. We never saw it. It made us sad, because even today, many years after the service lane and all the houses in Tyne and Godfrey Streets have been demolished, some people still say that Jessie and her cats can sometimes be seen on that spot on a Friday afternoon (Fortune 1996: 70, 94–95).

Jessie's Cats is one of many stories that gives depth – in the sense of human experience and memory – to the now deserted and scrub-covered townscape of District Six, an area of Cape Town from which people were forcibly removed under the Group Areas Act on account of their race or colour (Hall 1998, 2000 Chap. 7, 2001; Malan and Soudien 2002; Malan and van Heyningen 2001). Yet, 30–40 years after the first removals, much of the character of the District and of its former community remains, and with some prior knowledge of South Africa's recent past, this character *is* tangible (for example, Bell 1997). Of course, not everyone can or will experience this character in the same way: for insiders, such as former residents, memory will be the dominant factor; while outsiders will require more background to the events of the late 1960s and 1970s, and an awareness of the way of life of the District's former inhabitants, for the place to 'come alive' to anything like the same degree. This means that, for most visitors to Cape Town, interpretation is a necessary precursor to any site visit, and here the

J. Schofield, *Aftermath: Readings in the Archaeology of Recent Conflict*,
DOI: 10.1007/978-0-387-88521-6_3, © Springer Science+Business Media, LLC 2009

District Six Museum plays an important role, telling the story of the removals and acting as a catalyst for the numerous accounts and histories of the area that are now being told.

This chapter looks specifically at approaches to presenting and interpreting troubled pasts, primarily through the events of the apartheid era in South Africa, and explores how a combination of material culture, engaging museum displays, photographs and the use of narrative – preferably in the first person – is necessary if these events are to be presented as they really were, and not as some fabrication of truth, nor as a diluted, sanitised or unbalanced interpretation of past events (see for example Geiryn 1998; Linenthal 1995). The treatment of recent military and civil conflicts could easily be made rather cool and dispassionate to avoid controversy; it is argued here that in cases like District Six, engaging (or 'hot') interpretation is a necessity, with reconciliation an achievable objective (after Uzzell 1989; Ballantyne and Uzzell 1993; Uzzell and Ballantyne 1999). It should be stressed again that, in presenting this case study, the intention is only to give a particular spin to a story already well represented in the literature. For example, much has been written about the District's archaeology and architecture (Hall 1998; Malan and van Heyningen 2001; Fransen 1996), its social (for example Fortune 1996) and political history (Jeppie and Soudien 1990), its presentation as heritage (Ballantyne and Uzzell 1993) and its future (Malan and Soudien 2002). This chapter will use these sources, combined with personal experience as a first-time European visitor, to highlight the significance of the place, and the way it is remembered and presented, promoting it as an example of best practice for interpreting monuments and sites of conflict, injustice and atrocity. Some further examples of this same approach to heritage interpretation are also given.

'You Are Now in Fairyland': Interpreting Recent Events in District Six

A Brief History

In the words of a Museum brochure:

> District Six was an area of Cape Town at the foot of Table Mountain, near to the harbour and the city bowl. [It] was a cosmopolitan area. Priests, teachers, school children, prostitutes, families, politicians, midwives, gangsters, fishermen, pimps, merchants and artisans lived in the area. They came from all over the world and the different corners of South Africa and together created a rich mix of different cultures. They also introduced into South Africa a strong political tradition. The area was a seedbed of ideas and activities. Most of the people who lived in District Six were working class. They wanted to live close to the city, harbour and factories where they worked. Rich with memory, it was a place which has made a great contribution to the history and culture of Cape Town, and indeed to South Africa. As a result of the apartheid legislation, only the memories of District Six remain.

What happened to District Six under the apartheid regime is well documented (Hall 1998). Having been declared a Whites-only area under Proclamation 43 of the Group Areas Act (1966), virtually the entire district was physically erased from the map. Some 62,000 people had previously occupied the area according to government figures, which also indicated that three quarters of these were tenants, and all but 1,000 were classified as 'Coloured' (specifically of mixed descent) in the terms of the Population Registration Act. By 1978, some coloured families were still resident in the District, which by this time had become a rallying point for opposition to the forced removals that were taking place throughout South Africa. By 1984, the removals were complete. All that remained was the scar, separating Cape Town from its suburbs (Fig. 7): South Africa's Hiroshima, as one commentator described it; alternatively, 'the preserve of South Africa and all of humanity' (Nagra: personal communication).

In 1986, BP (Southern Africa) announced its intention to rebuild District Six as South Africa's first open residential area, once again attempting to impose policies on communities without consultation. BP's proposal further focused an already strong opposition and stimulated the formation of the Hands off District Six Campaign (HODS), an alliance of organisations and former residents which campaigned for the abolition of the Group Areas Act prior to any redevelopment. Abolition of the Act has since happened and in August 1997, a Land Court ruling gave the area back to former residents.

Fig. 7 Cleared ground at District Six, where all that remained were the churches. *Photo*: author

District Six today is an eloquent symbol of the policy of racial segregation that dominates South Africa's recent history, and of the sense of community which the apartheid regime sought to destroy. District Six was once heterogeneous and cohesive; there was no residential segregation between classes; and there existed a level of tolerance amongst people that could accommodate a range of religious and political beliefs (le Grange 1996: 8). This state of affairs was unacceptable to the apartheid regime and the clearances began. The District is now empty, 44 hectares of scrub which effectively hides the drama of the natural red earth, which 'bled' at the time of removals, but which does at least protect a rich and significant archaeological record documenting the history of the District's occupation and ultimately its clearance (Hall 1998). Furthermore, there remains a strong District Six community on the Cape Flats, and the plan now is to rebuild the District, returning some former residents alongside first-time occupants (this is discussed further below).

The District Six Museum

In 1989, ex-residents of District Six envisaged a museum to commemorate the area and honour the people who fought against the forced removals and Group Areas Act. On 10 December 1994, the District Six Museum opened with its first exhibition 'Streets – Retracing the Past'. The museum provides a space for the community to come together and share their experiences and retrace their memories. The District Six Museum is a reminder that forced removals must never happen again (Museum brochure, undated.).

The Museum is more than just a display. For a start it acts as the focus of a now dispersed community, and for this reason its location in the old Methodist Church on the edge of the District is particularly apposite. It was this church, for example, also called 'the freedom church', that took a stand against the injustices of the Group Areas Act and other apartheid legislation. It now serves as a meeting place, an educational resource and a point of contact. The Museum also has an important political role in the District's redevelopment, as well as acting as a conduit for narratives and personal accounts (publishing some of its own works and selling others), oral history, sound archives, and artefacts, such as those from archaeological excavations undertaken by the University of Cape Town in recent years. It was also closely involved in a public sculpture project in 1997, designed in part to reclaim the district.

The Museum is interactive. Former residents are encouraged to sign a cloth, which is later embroidered. Much of the ground floor is taken up with a map of the District, prior to the clearances, with the road names marked on. Here former residents sign their names and number the houses where they once lived, perhaps attaching brief comments or stories (Fig. 8).

The Museum also houses a photographic archive. When this was first shown publicly, it led to a celebration of life amongst former residents; singing, arguing and debating. And among the museum staff are former residents who will discuss

Fig. 8 The street map on the ground floor of the District Six Museum. *Photo*: author

the District with visitors, adding colour to an already engaging interpretive experience. What the Museum does not overtly do, however, is to show the horror of the removals. As many visitors have remarked, the power of the Museum lies in the fact that it has a celebratory air about it. There are no 'in your face bulldozers'; rather people are remembering themselves as a community, in a Museum which is essentially a homely place.

In 2000–01, the District Six Museum received comparatively few visitors; for example, only around 50 overseas tourists a day visited, mostly arranged through tour operators. Fewer still visit the District itself, probably because of concerns about personal security, even though the engaging and interactive interpretation which the Museum provides prepares visitors well for touring the District. To facilitate this, a leaflet has been produced by the Museum which provides a guided tour by Pastor Stan Abrahams, a former resident. One of the 30 points on the tour gives an impression of the whole:

Parkers was a corner shop which dealt in 'cash and credit'. Amongst other things, one could buy bread, paraffin for the primus stove and fish oil for frying fish. On hot Friday evenings, my brothers and I would push our way through the adults who were buying their weekly provisions to get to the soda fountain counter where we would spend our week's pocket money on bulls-eyes, almond rock or a koeksister (nd: np).

The tour takes in: the existing churches and mosques; the one terrace of cottages that survived the clearance and which today gives an impression of the District's original appearance; the cobbled streets, in many ways the centre of District life;

and the foundations of front steps from which people all over the District watched the passing scene.

To summarise then (after Hall 1998: np), between 1966 and the present, 'the raw scar of District Six was encrusted with a variety of meanings. For its former residents, it was marked ground, the geography of dispossession and dispersal. For the apartheid government it stood for White entitlement and the principle of separate development. For reformist business and municipal interest, the land was an opportunity for investment and economic development'. It remains now to consider the future of the District, and specifically the role its physical remains play in presenting and interpreting its troubled past.

The Future

Proposals to redevelop District Six have been under discussion for some years. It is a controversial matter and one which the local community will have to resolve with politicians and city planners if a mutually acceptable solution is to be found. From a conservation perspective, it is important that the character of the District and at least some of its physical traces are retained to work alongside the Museum in interpreting the past, for three main reasons:

First, for the sense of belonging such areas provide for their former inhabitants. The Museum at District Six for example has served to galvanise a community that was scattered amongst the townships in the years following the passing of the Group Areas Act, while the 44 hectares of empty ground (excepting the churches and mosques that remain) has, throughout the apartheid years and beyond, acted as a daily reminder of the removals, to Capetonians and visitors alike.

Second is the 'lessons from history' argument that the social injustice of the forced removals must be kept in the past. There is also the hope that lessons from South Africa will eventually attain wider geographical and geopolitical significance and influence.

Third is that increasingly people want to know about the recent past, and in particular about the momentous events of the twentieth century. What happened in South Africa under the apartheid regime constitutes a major episode in recent world history, and District Six tells that story arguably better than anywhere else (but see below, for reference to Robben Island).

What is finally agreed will need to be sustainable in the long term, and for this reason alone the strategic location of the vacant land, the size of the area and the increasing need for affordable inner-city housing suggest that a significant amount of development will be necessary; this is perhaps appropriate in the circumstances. Re-housing those forcibly removed sounds attractive, but not all individuals removed can (or perhaps even want to) return, and new developments such as the Technikon – built originally for White students only – cannot simply be removed. But as le Grange (1996: 15) has sensibly argued, District Six can still be used as a model for how to address the wrongdoings of the past and as a way to begin healing a divided city. Of course this would require the participation of the affected community and the concerted political will of government to deal sensitively with the planning and implementation of a reconstruction programme. Three specific aspects of this model can be identified:

First, it is important that the future development of the District draws upon the urban planning traditions of its past. For example, the fine-grained street network; the mixed land-use development; a mix of housing types to ensure social heterogeneity; the street as community space; and the population density that shaped the area and which can be reinterpreted and adapted to serve contemporary requirements (after le Grange 1996). The surviving churches and mosques could serve as foci within a redesigned District Six, with one of them permanently housing the Museum.

Second, views and vistas will be important, particularly for former residents revisiting the District. To this end, le Grange (1996) produced designs to retain as open space an area either side of the sloping and cobbled Horsley Street which uses mounds of rubble to obscure the foreground, yet showing glimpses of the city, a view that residents would have had. This area also includes the site of one of the three excavations undertaken in the District by the University of Cape Town; artefacts from these excavations could remain on view at the Museum.

Third, places of memory should be (and in fact are being) considered, to serve for example as areas for quiet reflection and play. In 1993, the District Six Museum Foundation called a public meeting to get sanction from the community to set aside land in District Six as 'memorial parks'. This remains an option at the time of writing and would be important for many reasons, such as allowing easy (including virtual) access to the District for those who cannot or choose not to return as residents. As we have seen, the front steps of houses, from which residents 'watched the passing scene', survive in some areas, along with original cobbling. In terms of presenting and remembering past events, these steps and cobbling are arguably the most meaningful of all material remains surviving within the District and would be an important component of such 'memorial parks'.

In summary, District Six is an evocative and an important place, both for former residents and visitors. For those who understand the significance, its atmosphere is tangible, obviously so for former residents but for visitors too. What happens to the District matters to all these people, but most significantly of course it matters to the future generations who will visit and interpret it. For those future visitors, the steps, the cobbles and the bare red earth may be the most powerfully symbolic of all that remains, and for that reason alone, the arguments for their retention are compelling.

District Six is not unique in these terms and a few examples follow where comparison with aspects of presentation and interpretation at District Six demonstrates and confirms the strengths of this approach.

'Don't Forget Us': Other Troubled Pasts

Wartime Monuments in England

As we have seen already (Chap. 1), English Heritage has since 1995 been undertaking a national review of England's recent monuments of war, developing the understanding necessary to secure their future management (English Heritage 1998). As a result of this work, some sites – examples of the typical and commonplace as well as the rare – will be afforded statutory protection through scheduling, some will be listed, while others will be managed locally through the development control process (but see Chap. 1 also for mention of recent changes to the heritage protection system in England). Many of these sites are accessible to a public who are increasingly aware of and interested in the fabric of war – popular books, television programmes and museums have ensured that. And an approach to interpretation and presentation, which engages the visitor and which is in keeping with the site's original function, is now fairly typical. Orford Ness, for example, a Cold War experimental site on the Suffolk coast, is presented to visitors through a 'philosophy of non-intervention'. An extract from the guide book explains that this philosophy: 'stems from a need to protect the features and geomorphological value… as well as its aura of mystery. … The main structures of the buildings and their impact on the landscape should survive for many years; and their symbolic value will thus be maintained for future generations'.

Orford Ness is thus presented in a way that is compatible with its role in experimentation and atomic weapons testing. Similarly, Cold War control bunkers, such as that advertised on road signs for miles around as 'The Secret Bunker' at Mistley (Essex), are typically presented as cold, grey structures which visitors depart feeling suitably chilled, physically and psychologically. At Mistley, the monotony and silence of empty corridors and bare rooms, mostly underground, is broken by three cinemas, each showing public information films of the time. In this atmosphere, the '3-min warning' is a dramatic interruption.

Preserving such sites serves several purposes, one of which is the opportunity they provide for exploring, experiencing, interpreting and deconstructing the recent and contemporary pasts. The statutory protection of Second World War anti-aircraft gunsites is an example of this. Over 2,000 heavy anti-aircraft gunsites were built in England in 1939–45, their distribution confined mainly in the east and south, and around cities and industrial centres – in other words, the areas or places most vulnerable to attack (Dobinson 2001). From studying modern aerial photographs of known sites, it has been established how few of these sites survive (Anderton and Schofield 1999), and how rare are those examples where the layout of the site – with its gun emplacements and control building and the domestic sites – provides a visual impression of scale and function. Plans and photographs are one thing, but for visitors wanting to appreciate the site's layout, the spacing of buildings, and their configuration and alignment, to experience the 'ghosts of place' (Bell 1997; Edensor 2005), the survival of structural remains including the original road layout are necessary, ideally in an environment virtually unchanged since 1945. Although some examples of incomplete gunsites will be protected (in view of their overall rarity as a monument class), complete examples that enable interpretation are a priority.

Unfortunately, the urgency with which these sites must be protected and the speed of the national review of which this example formed a part (the Monuments Protection Programme [MPP] – see English Heritage 2000 for further details) meant that information from those serving on these sites could not easily be included in site assessments, or in the resulting documentation, unless already available in local records or in published form. The value of such testimony as a source of information is recognised however. At Brixham (Torbay) for example, an emergency coastal battery survives well, with its gun emplacements and associated buildings. Some of these structures are now protected while the wider site has a conservation area designation. But what really gives this site colour, what brings it to life for those that visit, is the fact that several of those who served on the Battery live locally and keep an eye on the place. An interpretation booklet and panels have been provided, lectures are given locally about the site, and most impressively, veterans approach visitors to the Battery with the offer of guided tours.

Blitz Experiences

Wartime monuments enable Second World War and Cold War sites to be experienced by a public who are increasingly knowledgeable of and interested in the material culture of these recent historical events. But as with District Six and Oradour (below), engaging museum displays have a complementary and significant role. The experience of the Blitz, brought to life to varying degrees by the Blitz Experience at the Imperial War Museum, and another at the Winston Churchill Museum, both in London (and critically reviewed by Noakes 1997), for example provide a focus for exploring landscapes of the Blitz in contemporary London (Holmes 1997). To take the last first, the display at the Winston Churchill Museum encourages the visitor to understand the Blitz by 'experiencing it'; to share the wartime experience, to 'see it, feel it, breathe it… be part of those momentous days'. As Noakes describes it (ibid.: 96–97):

> Descending in a rickety lift, the visitor emerges into a reconstruction of a tube shelter, where she or he can sit on original bales of wartime blankets, to watch a collage of wartime newsreels.… Emerging from the Tube shelter, the visitor next walks along a corridor lined with photos of London during the Blitz and newspaper headlines of the time. At the end of this corridor the visitor can choose to enter an Anderson shelter, where she or he can listen to recorded sounds of an air raid, look at an exhibition, or pass on to the centrepiece of the museum, the 'Blitz Experience'.

The Imperial War Museum's 'Blitz Experience' is rather different:

Visitors are ushered into it by a guide, entering through a small dark doorway to find them-selves in a reconstruction of a London brick-built shelter. The shelterers are urged on into the shelter by the taped voice of George, a local air-raid warden. As the shelter fills more voices appear on the tape, all with strong London accents. Some talk about their day whilst others complain of lack of sleep. As the bombs begin to fall, George leads them in a hearty rendition of 'Roll Out the Barrel'. As the bombs get closer George's daughter Val becomes hysterical, her screams gradually drowning out the singing. A bomb drops uncomfortably close and the shelter reverberates. Everything goes quiet.

The shelterers are then helped outside by the museum guide, whose flashlight plays around the devastated street that they are now standing in. In front of them lies an upturned pram, its front wheel still spinning... As the shelterers leave blitzed London to become museum visitors once more, their last experience of the Blitz is George's fading voice saying 'Don't forget us' (ibid.: 95–96).

There are common factors here. Both experiences involve damage to property not people. Emerging from the experience at the Winston Churchill Museum, what may at first be thought to be bodies are, on closer inspection, mannequins from a bombed shop, though the initial impression may be deliberate. Also, both experi-ences are of large communal shelters, even though these accommodated only a small percentage of London's population. As Noakes puts it (ibid.): the experiences represent a sanitized version of a minority experience presented as a majority expe-rience, and the display at the Winston Churchill Museum bears little resemblance to the Tube shelter recalled by a former shelterer in Calder's *The People's War* (1969: 183) who described a place where, 'the stench was frightful, urine and excrement mixed with strong carbolic, sweat and dirty, unwashed humanity'. Yet despite obvious limitations in telling the typical Blitz experience precisely as it was, these are engaging displays. There are personal accounts to be read, photographs to be seen, and – not too far away from either Museum – bombsites to be visited, such as the ruined churches of St. Mary Aldermanbury, St. Dunstan-in-the-East and Christ Church. We will return to these churches in Chap. 3.

Wartime Atrocities

Sites of wartime atrocities, and notably Holocaust sites, present particular problems for presenting and interpreting past events, and Gilbert's *Holocaust Journey* (1998), in which he describes a field course for MA students studying the Holocaust, out-lines some of these difficulties as well as demonstrating the effects an engaging display can have on its visitors. Their visit to the lakeside villa, where on 20 January 1942 Reinhard Heydrich introduced the 'Final Solution' to ministerial bureaucrats, is described thus:

In what is believed to have been the actual room in which the Wannsee Conference took place, with its tall windows looking out over the patio and lawn down to the lake, there is a stillness. We walk into the room, through it, round it and then out of it, as if it must not be disturbed. It is as if the voices of those who spoke here, and the heads of those who nodded their agreement here, must not be alerted to our presence. One feels a palpable sense of the presence of evil (ibid.: 47).

Gilbert goes on to describe how his students were 'deeply affected by the visit to the Wannsee', how the meeting had been so clinical and how the interpretation of events now presented this in a direct and unencumbered way. As one student put it: 'You don't get lost. You don't get bogged down. It's all depicted in a nutshell. Very comprehensive. It is a credit to the authorities that they have decided on this place – of all places – to have this mind-boggling exhibition' (ibid.: 50).

Another stop on their 'journey' was Prague, where they visited the Orthodox Cathedral Church of St. Cyril and St. Methodius, where Heydrich's assassins were trapped and killed in June 1942. Although not explicitly a Holocaust site, the interpretation of events provided for visitors is worth recounting:

> We enter the crypt.... There is an exhibition provided by the Imperial War Museum in London, as well as a film. The story is tragic; [after their betrayal] the trapped men [Czechs, trained in Britain] barricaded themselves in the crypt and tried to dig their way through the brickwork into the sewers. The hole they dug penetrated six feet into the brickwork but they could get no further. The Germans pumped water into the crypt. When this failed they pumped in smoke. Finally they burst in. The men refused to surrender and were killed in the crypt. The hole they were digging is preserved as a memorial; it is a shattering site.
>
> The film starts. It is a dramatic reconstruction, and it is a strange sensation to be standing in the room which is being portrayed in the film. After the silence and sombre nature of the crypt, however, the noise of guns firing is jarring. Most of us drift out before the end. There is something unreal, but also unnerving, about the reconstruction (ibid.: 64).

Gilbert gives many other examples and tracing his students' reactions to how the past is presented at the Wannsee, and particularly at sites like Auschwitz and Belzec, provides an interesting dimension to the 'journey'. The particular point, that it is his 'readings' of contemporary accounts that most influences their reaction, is noteworthy in the context of the other examples presented here. As Gilbert said about the effect of his 'readings' at the death camp of Sobibor (1998: 251): 'It is difficult [to read the passages I have prepared]. Even words written by survivors seem to intrude on the awfulness of the place. And yet, without these words, the awfulness is somehow diminished.'

Contemporary photographs are a further and powerful medium in interpreting atrocities of the Holocaust. At Auschwitz-Birkenau, photographs set up on two small exhibition boards form part of a collection known as the *Auschwitz Album*. These include the only known photographs of people arriving at Birkenau, 'waiting, bewildered and uncertain, before being taken to their deaths' (Gilbert 1998: 160–161). Gilbert tells the story of this collection, including the circumstances of its discovery:

> The photographs had been taken on a single day (quite illegally), probably by SS Second-Lieutenant Bernhard Walter, Director of the Identification Service at Auschwitz.... The pictures were put in an album with neatly inscribed introductory captions, the first of which read: 'Resettlement of the Jews from Hungary'. Several months later, it would seem that a guard at Auschwitz named Heinz (his surname is unknown) sent the album to a guard at

Nordhausen, perhaps his girlfriend, and probably someone who had earlier served with him at Auschwitz. He inscribed the album: 'As a remembrance of your dear, unforgettable, faithful, Heinz'.

[…]

The album was found in Nordhausen concentration camp on the day of liberation in May 1945 by 18-year-old Lili Jacob… who had earlier been deported to Auschwitz from the Beregszasz ghetto. She fainted when she found the album: for among the 193 photographs was one of the rabbi, Rabbi Weiss, who had married her parents. On continuing to look through the album, Lili Jacob found photographs of two of her five brothers, 11-year-old Zril and nine-year-old Zeilek. They had both been gassed shortly after the picture was taken, as had her parents, her other three brothers, her grandparents, and her aunt Taba and her five children (who also appear in the album) (ibid.).

As Gilbert states (ibid.: 161), these are terrible pictures, since we know the fate of those standing about, sitting with their bundles or walking along the fence. But they only record a single day out of more than 800 on which deportees arrived here at one camp of the many that existed. Standing there today, looking at the photographs, and looking around, one can almost match up the trees with those in the photographs, giving an immediacy that only an in situ presentation could achieve.

Finally, at Oradour-sur-Glane in western France, a company from Das Reich armoured division killed 642 people on 10 June 1944, among them 244 women and 193 children and babies. Most were locked into the church, which was then burnt down. Every year half a million people visit the town, where burnt-out houses, cars and public buildings have been maintained as they were left in 1944 (Uzzell 1989), following the decision not to rebuild the village but rather construct a new settlement on the outskirts. Until a few years ago, a tour guide, related to one of those killed, took visitors around the town, explaining in detail the events that took place there (Uzzell and Ballantyne 1999: 156); tourists also visited the small museum, where personal effects can be viewed. On 9 July 1999 Jacques Chirac opened a vast £6 million underground war crimes centre – Le Centre de la Memoire – in the village. The discrete architecture – a 10,000 sq ft crypt whose low entrance is through a giant mirror reflecting the rural valley overlooked by Oradour – underlines (in the words of those responsible): 'the contrast between the gentleness of the valley and the sombre aspect of the ruins.' More significantly, the new centre now provides the interpretive context for atrocities conducted over a longer period embracing much of the twentieth century, in a place where the impact of a single such event can still be experienced by visitors. In this sense, the co-location of Le Centre de la Memoire and the burnt out remains at Oradour will convey a powerful message to visitors as memory of the atrocity fades.

Robben Island

Returning to South Africa, another significant monument to past troubles, and specifically to the apartheid era, is Robben Island, whose international importance is reflected in its World Heritage Site status. This was the place where, after serving as a convict station, farm and leper hospital, and after fortification during World War II, it became known the world over as a place of brutality, harshness and a symbol of human rights abuses under the apartheid regime. Famously Nelson Mandela was imprisoned on the island for 27 years (Clark 2002; Smith 1997).

Here the experience of visiting the island – now the Robben Island Museum – is engaging and highly charged; the story is told as it was and importantly by those who were there. The focus of the visit is not surprisingly the prison, in which photography is forbidden, and where the tour guides are one-time political prisoners, some of whom were held on the island and whose tour is – inevitably – an intensely personal account, if not of Robben Island then of the experience of political prisoners elsewhere in South Africa. It is not just the prison buildings that convey this experience; a visit to the island also includes the quarries where prisoners laboured, and suffered eye damage from the sun's reflection off the limestone, and the cells where some prisoners – like Robert Sobukwe – were kept in solitary confinement over many years. This insight into the island's landscape beyond the confines of the prison, its obvious remoteness, and particularly sharing the tantalising glimpses prisoners had of Table Mountain, give a clear impression of a further perhaps more obvious dimension of the apartheid era, telling another part of the same story that is presented so effectively in District Six.

Tourjeman Post Museum, Jerusalem

To contrast with the success of presenting and interpreting the past at District Six and Oradour amongst other places, is Jerusalem, and a programme for the renewal of the Tourjeman Post Museum following the peace process of the early 1990s. Here the difficulties of dealing with recent or ongoing troubles are exemplified, demonstrating that, despite the best of intentions, some situations will be too hot to handle, for the moment at least.

In what was originally a museum devoted to Israeli heroism, the intention was to establish a 'museum of co-existence', an establishment where the narratives of the city's Palestinian and Jewish parts could be set out together, side by side (Ben-Ze'ev and Ben-Ari 1996). Within Jerusalem, the division between the Arabs and the Jews characterises the city. These two national groups are spatially divided, and are separated in terms of education and employment, and it was this separation that eventually led to the project's failure. As Ben Ze-ev and Ben-Ari state (ibid.):

> [it was in the] negotiations, struggles and discussions that the predicaments of creating the museum emerged in their full force. We emphasise that despite the expectation that a museum of 'co-existence' could be established, in retrospect this was but an illusion. Politics imposed itself to forestall such an opportunity.

Conclusion

The examples used in this chapter cover a range of topics, social conditions, political contexts and – of course – spatial and temporal diversity. The common thread however has been the presentation and interpretation of past troubles in a particular way and with the emphasis on human experience and its material manifestations. This is partly demand-led. There is considerable interest in the materiality and monumentality of the recent past, just as there is fascination with conflict, military especially, but also civil unrest and social injustice. There may need to be a cooling off period before such things are presented as heritage, but that need not necessarily be long. It is interesting to note that in Berlin for example, where the Wall was seen as a symbol of division in a once united city, it took only ten years, from the Wall's demolition and a recognition after the initial destructive phase of the need to preserve short sections, for a debate to start about its 'reconstruction' in some areas. So the point is not so much about when, but what and how, and it is on these issues that this chapter has laid emphasis.

To summarise some key points from the examples quoted, they have emphasised the need for (re)presentations of the past to be:

Accurate: Displays should aim to tell the story as it was, and not some sanitised or diluted version or fabrication of the truth; stories can of course be tailored for younger visitors, and some sensitive information can be effectively hidden in the more technical guidebooks or 'top-shelf' display facilities, if that is considered appropriate. A constructivist approach is generally favoured however, encouraging visitors to think for themselves, based on the evidence presented. Of course this desire for accuracy in no way prevents or discourages the production of alternative histories. Equally, and on occasion, controversy – either about what happened or the implications of certain events or actions – will prevent a consensus of opinion, and in such cases displays may never materialise. This proved to be the case with the now well-documented disagreements over the Smithsonian's display about Hiroshima (see Gieryn 1998; Perkins 1999; Linenthal 1995 for example). Alternative views and interpretations of the past should be a significant component of displays and exhibits.

Facilitating: A sense of the place and its distinctive qualities and characteristics are not given, and therefore cannot necessarily be passed on only by interpretation. It is created by individuals, and the aim of displays should be to give people the means to develop their own appreciation of significance. Interpretation should facilitate. There is also a need for sense of place to be owned and to grow out of individual experience, needs and perceptions. The sense of discovery is vital. Residents and visitors – and scholars and interpreters – should jointly participate and share their perceptions (Graham Fairclough: personal communication).

Engaging: Presenting troubled pasts will be most effectively achieved by emphasising human experience. This can be achieved in different ways and in this chapter we have seen several: the value of written accounts, through Jessie's Cats for example; the use as guides of former prisoners at Robben Island and former residents at Oradour-sur-Glane and District Six; audio-visual techniques to tell the story of Heydrich's assassins, and the simple and uncomplicated use of photographs at Auschwitz-Birkenau; the direct involvement of veterans in managing Brixham Battery; and the use of fictional characters, represented by Val and her growing hysteria in the underground shelter in the Imperial War Museum's Blitz Experience. Another notable approach is for museum visitors to take on the identity of contemporary characters. This has proved to be both popular and successful at the In Flanders Fields Museum, Ypres, and the Holocaust Museum at Washington DC (see also Chap. 3).

Respectful of past events: It is important that the correct balance is struck between providing a tourist 'attraction' and preserving the character of the place one is presenting to visitors. Plans to develop Auschwitz-Birkenau are an example: proposals include enlarging the parking area, building a small by-pass around the main gate and a reception area opposite it, and converting the sauna (one of the few buildings to survive intact, excepting the huts) into a museum. Gilbert's reaction to these proposals was one of bewilderment, and in his view 'no doubt admirable from the museum curator's perspective, but incongruous after what [he and his students had] just seen' (1998: 173).

Signposts and symbols: It is important that the engaging, accurate museum displays aren't left alone in presenting troubled pasts (Schofield 1998), though they do of course have an important part to play. While these displays do contain touchstones through which visitors can gain insight into ordinary lives and personal experiences, the places themselves can be more powerful still in achieving these objectives, both in terms of the atmosphere or character of the place, and in its material remains. The open ground at District Six has extraordinary character, and has retained its sense of place and identity, a point reinforced on the 1997 Heritage Day holiday, when several thousand people 'reclaimed' District Six, to look at art and listen to music (Hall 1998). The ground also has the potential to be powerfully symbolic, the bare red earth acting as a reminder of the physical act of forced removal.

The general point here is that sites (in the sense of monuments and places) documenting troubled pasts, and especially those which involved human suffering, should attempt to bring the place alive for visitors, and most of the examples given in this chapter achieve that, provided the visitor is suitably informed. To talk of ghosts may seem unscientific, but it is a valid point. A final example from Gilbert's *Holocaust Journey* illustrates this, quoting from the *Jewish Chronicle* John Izbicki's account of the re-inauguration of the restored Orianenburger Strasse Synagogue in Berlin on 7 May 1995, where he had prayed as a young boy before emigrating to Britain in 1939:

> We all sat outside, on the ground where two-thirds of the original building once stood. This empty space, where the main hall of the synagogue used to be, is to be left as a lasting scar of history. It is the remaining one-third that has been transformed into a museum, a place for researchers to come and study the history of German Jewry. The roof of that one-third is adorned now, as it was before, with two golden cupolas that shine like beacons across the Berlin skyline. As I listened to the speeches of eminent personalities and looked up at the windows of the restored building, I thought I saw – and certainly felt – the presence of so many others who had once prayed there. (Quoted in Gilbert 1998: 36)

To return to the opening quotation, it is argued here that for the interpretation of past troubles to be affective (and thus effective), stories should be presented which have characters to whose lives the visitor can relate. With the recent past we have those characters in abundance, real people whose lives and activities, whose sacrifices, heroism or evil deeds, give the stories a strength which could never be generated in fiction. We should make the most of this in our presentations. This is exemplified by District Six, where the history of the area and the sense of community that continues to exist here make it an important heritage site. With the museum displays and the literature available, it illustrates the way recent troubles can be presented to a public who increasingly want to know. At the time of writing, the future of the District has yet to be resolved, but given strength of opinion and the depth of commitment locally, a sensible and sustainable solution will surely be found serving as an example to others, both in the new South Africa and beyond.

Chapter 3
Monuments and the Memories of War

As we have seen, over the past two decades, twentieth-century defences, fortifications and experimental and military production sites have become an accepted part of the cultural heritage, in Britain as elsewhere (Dobinson et al. 1997; English Heritage 1998; Cocroft 2000). For heritage managers, planners, archaeologists and historians this has meant learning a new vocabulary, and intricate typologies for such things as anti-invasion defences of the Second World War, radar establishments and coast artillery. It has also meant the need for some grounding in scientific principles, particularly relevant for studying radar and Cold War facilities, as well as artillery and ballistics. It has meant developing a theoretical framework for interpreting and managing these contested sites. It has meant new conservation challenges, such as the practical measures for prolonging the lives of concrete structures designed to last only 'for the duration'. It has meant developing an approach to interpretation that balances the various needs of cultural tourism with the response these sites often evoke (Chaps. 1 and 2). And – significantly – it has meant close co-operation between professionals and those amateur archaeologists and historians who have been responsible for much valuable groundwork over the last 30 years or so (Wills 1985; Morris 1998).

But why has this willingness to embrace recent military heritage been taken up with such enthusiasm and alacrity? What are the motivations for conserving what are often ugly, functional and unstable buildings? And why in particular is it important that some of the buildings and structures remain when publicly available records are known to exist, and where recording prior to demolition provides a lasting archaeological record for reference and research? These related issues around the subject of conservation form the basis of this chapter, with the emphasis on two specific aspects of the Second World War in England: the air war and the urban Blitz of 1940–41. It is argued that monuments relating to these episodes or aspects of the conflict have a particular role as 'living memorials', and this reason for their retention will form the basis of discussion. (Other specific events, notably D-Day and the Battle of Britain form the subject of later chapters in the section Landcapes of Events.)

J. Schofield, *Aftermath: Readings in the Archaeology of Recent Conflict*,
DOI: 10.1007/978-0-387-88521-6_4, © Springer Science+Business Media, LLC 2009

Motivations

In England, as elsewhere, work has been underway for some years with the aim of understanding recent military remains sufficiently to provide a credible assessment of their importance and to inform options for their future management (Dobinson et al. 1997; English Heritage 1998 – see Chap. 1). This assessment operates at three levels: in a holistic sense, for the subject matter as a whole (that is, twentieth-century military sites); for individual monument classes (such as anti-aircraft gunsites, cf. Chap. 2); and – within each of those classes – for the sites themselves. The first of these levels concerns the values attached to this category of site by society; the second and third are about importance, and how we (as individuals, or the heritage sector) might choose between particular sites and types of site. These last two stages will be subject to further scrutiny in the second half of this chapter.

Beginning at the most general level, several considerations are relevant. A concern often cited is that retaining the fabric of recent wars only serves to delay the healing process, and prolong what are often bitter historical tensions and rivalries. Virilio (1994), for example, noted how German fortifications along the French coast continue to provoke responses of hostility (several bunkers sporting hostile graffiti), bewilderment (passers-by rarely recognising the bunkers as archaeological), hatred and vengeance. These last two are perhaps the most interesting of these responses: many installations were destroyed when France was liberated – basements were filled with munitions and blown-up, the explosions 'delighting the countryside's inhabitants as in a summary execution' (1994: 13). In studying these sites during the 1970s, Virilio recalls being told by local inhabitants how they frightened them and called back too many bad memories. '[They provoked] fantasies too, because the reality of the German occupation was elsewhere, most often in banal administrative lodgings for the Gestapo; but the blockhouses were the symbols of soldiery' (ibid.).

Then there is the extent to which the past provides 'lessons for the future'; 'the same mistakes will never be made again'. Yet they are repeated, and they continue to be repeated today in the many internal conflicts and ethnic wars to emerge from the ending of the Cold War. The horrors of the Holocaust are well documented, and some of the key sites are preserved (one – Auschwitz-Birkenau – is a World Heritage Site). Films such as *Schindler's List* reveal the atrocities to a new generation. Yet as cinema audiences recoiled from the gruesome scenes in a Hollywood movie, the Muslim population of Bosnia was reliving them (Rupesinghe 1998: 1). In this particular instance, therefore, the preservation of concentration camps, combined with films, books and engaging museum displays such as at the Holocaust Museum in Washington D.C. (Weinberg and Elieli 1995), appears not to have prevented repetition.

Despite these arguments, however, there is a consensus that some recent military sites should be preserved for the benefit of this and future generations, and several reasons for this are generally given. Prominent among these is that the materiality of war crystallises military thought, as well as providing reference points or landmarks to the totalitarian nature of war in space and myth (Virilio and Lotringer

1997: 10). It is a part of the heritage which tells a fascinating story and as such provides a significant cultural and educational resource, illustrating the key events of the twentieth century, Hobsbawm's 'age of extremes' (1995). Even the humble pillbox can provide an opening to the experiences of war: the imminence of invasion; the scale and speed of the construction task; the nature and mobility of 'total' mechanised warfare; 'children's playful warring... after the real warring' (Virilio 1994: 15). Purpose-built slips, from which troops embarked for D-Day, provide a focus for commemoration and remembrance services (Chap. 11). Bomb sites like Coventry Cathedral provide a physical reminder of the scale of civic destruction, as well as a context for the act of personal and collective remembrance and a symbol of post-war regeneration. These values exist in the contrast the cathedral provides between the powerfully destructive forces of war, and the tranquility of enclosed spaces in the modern townscape. And control towers, which often survive as ruins on desolate airfields, stand as powerfully iconic structures of the air war, providing a focus for the memories of veterans who continue to return to airfields on which they served.

As a part of the heritage, therefore, war – and especially the Second World War – has educational and emotive value that gives its materiel culture particular resonance. And, though not unique to this subject matter, there is the additional benefit of personal testimony. The availability (and accessibility) of military records is conducive to detailed studies, not only of the famous (see for example Morris 1995), but also of family members and friends who served, in whatever capacity. In addition to these personal testimonies – some in the form of diaries written at the time, some as reflective accounts – official record books and other military records place individuals in certain contexts at specific times; while names on memorials provide an immediate and personal reference point to past events and lives (Chippindale 1997: 509). This is a past with people whose lives can be investigated through documents, testimony and places. And this potential to move beyond the impersonal has been realised in increasingly engaging ways. As we saw at the end of Chap. 2, the Holocaust museum (Washington D.C.) 'heightens empathy by making the horrific legacy intensely personal: each visitor wears the identity tag of a specific victim, a ghostly companion whose persona the visitor adopts and whose fate is disclosed, with haunting impact, at the tour's end' (Lowenthal 1996: 142). Similarly, at the In Flanders Fields Museum at Ypres, visitors are assigned real historical characters. As they pass through the museum, they learn the destiny of this person. As the curator explains, 'The museum is a place of encounter – an encounter with people of 1914–18 – not only soldiers, but doctors, nurses, writers, artists and children'.

In general terms, therefore, these sites and monuments of war are valued, both by those involved in the events being recalled, and their memories, but for this and (potentially) future generations too. The remains have cultural and educational benefit, as well as economic potential, if marketed effectively.

From this discussion of general principles stem two specific examples of the approach taken in England towards sites associated with offensive and counter-offensive operations in the Second World War. These site types present a particular challenge for conservation, being those most directly associated with a combination of human suffering and loss of life and, in the case of the Blitz, with the destruction of cultural property and civilian losses. By briefly describing the materiality of these aspects of warfare, the role the sites play in contemporary society will be assessed.

Two Examples

Bomb Sites

The aerial bombing of civilian targets has its origins in the First World War, but is better known as a characterising feature of the Second, which reached a climax with the dropping of atomic bombs on Hiroshima and Nagasaki to bring the Pacific War to a close. Between 1939–45, British and German towns and cities were subject to often intense aerial bombardment: in Britain, for example, over 60,000 civilians were killed and more than 86,000 seriously injured as a result of aerial bombing alone. Many town and city centres were badly damaged, requiring planning and regeneration in the immediate post-war years.

In England, the assessment of some site types associated with aerial bombing has been undertaken by English Heritage (see Chap. 1, Table 1). Urban areas affected by damage have been assessed in the past, but only in terms of the potential for earlier surviving buried remains and the opportunity these provide for archaeo-logical research in its conventional sense i.e. area-based excavations, usually of the Roman and medieval cities. But attention has now turned to assessing these remains as *bomb sites*, while exploring conservation options for this part of the historic environment. That this wasn't undertaken before 1999 is surprising given that argu-ments about conservation and regeneration at various notable sites were well rehearsed during the war and in the immediate post-war years. St. Michael's Cathedral, Coventry was the only one of Britain's (then) 59 cathedrals to suffer badly from bomb damage. Its surviving shell became, 'a symbol of the wastefulness of war' and the decision was taken to retain it next to a new cathedral. Similarly, the church of Holy Trinity, Plymouth still stands as a ruin. After it was bombed, the church authorities decided not to rebuild or to remove the ruin as part of the city's redevelopment; it was finally purchased in 1957 by the Plymouth Corporation, 'to be preserved as a memorial to all the civilians of the city who lost their lives during the war from enemy air attacks'. This approach to the conservation of ruined struc-tures was in fact a global response in the immediate post-war years. In Berlin, the Kaiser-Wilhelm Gedachtnig Kirche was levelled by bombing raids, except for the western spire and portal which, after some debate, were preserved as a 'memorial to peace'. In Hiroshima, the Atomic Bomb Dome, the former Prefectural Industrial Promotion Hall, is the only ruin left from the atomic bomb and forms the centre-piece of the Peace Memorial Park (Beazley 2007). In fact, only days after the bombing of Hiroshima newspapers stated that '(all) ruins should be left as memorials', but only on the false premise that the city 'could not now be occupied for 70 years' (Toyofumi1994: 39).

As well as select bomb sites of the Second World War, more recent examples of bomb damage have also entered the debate about the relative merits of: conservation as ruins and memorials; clearance for urban regeneration; or rebuilding. This was one of the issues debated at a World Archaeological Congress Inter-Congress on 'The Destruction and Conservation of Cultural Property', Croatia, May 1998 (Layton et al. 2001). For example, clearance for urban rebuilding and regeneration (with archaeological conditions) has occurred in Beirut, although 308 historic structures will be restored, chosen because of their architectural, historic or religious significance (Raschka 1996).

Usually, where sites of obvious and uncontested heritage merit are concerned, rebuilding is favoured, with the sites' symbolic value often enhanced as a result. Some churches and civic structures were restored in England after the Second World War for example, often to act as the centrepieces of visionary post-war redevelopments (as at St. Giles Cripplegate, London). Following the launch of the Northern Ireland Assembly (1 July 1998), ten Catholic churches in the Province were fire-bombed – talk was immediately of rebuilding, 'whatever the cost'. St. Ethelburga, Bishopsgate is now known as much as the church rebuilt after being partially destroyed by an IRA bomb in 1993, as it is the smallest church in London. After rebuilding, its role is as a 'Centre for Peace' (The Ecclesiological Society 1994). Finally, the Ottoman bridge at Mostar, shelled and destroyed by Croat militiamen, symbolised the idea of a multicultural Bosnia, but was targeted as part of the destruction of the identity of an entire people. Rebuilding the bridge was controversial, many regarding it as an empty gesture, merely recreating a checkpoint between Croatian and Bosnian-controlled parts of the city: a road to nowhere. But a poll of residents in eastern Mostar – mainly Muslims – felt the bridge should be the last monument of that conflict reconstructed, 'for shame, for mourning', as one resident put it (Dodds 1998).

Attitudes to bomb damage will vary therefore, depending on social, economic, cultural and political arguments. However, in England, and in the present climate of sustainability and urban regeneration, it is those structures which remain as *ruins* (and which to some are 'eyesores') that present a particular conservation dilemma between heritage and economic and social interests. Two points are of particular relevance here:

First, bomb sites may have been significant structures prior to being damaged, as well as now having value as memorials. Furthermore, there may be sites whose main heritage interest rests in their being situated within areas of nationally important buried (and earlier) deposits. On the other hand are sites where neither of these conventional heritage values applies. What all these sites have in common, however, is the degree to which they have accumulated a *symbolic* value over the last 55–60 years, some almost instantaneously (like the Atomic Bomb Dome), and some more gradually. Some of these sites have international significance in these terms (Coventry, Berlin, Hiroshima), while knowledge of others is more locally based, such as the bombed cinema in Kingston-upon-Hull (Fig. 9).

This National Picture Theatre only came to public knowledge following a local campaign to see the ruin converted to a memorial garden. While no one died in the

Fig. 9 The National Picture Theatre, Kingston-upon-Hull: bombed during the Second World War and still present in the urban landscape as a bomb site and a little-known local landmark. *Photo*: author

bombing, it does stand as the last obvious and tangible reminder of the Blitz in one of the worst-hit cities in Britain.

Given the extent to which towns and cities subjected to aerial bombing during the Second World War were planned and redeveloped in the post-war years, it is not surprising that so few bomb sites now stand as ruins. Many town centres were cleared and redeveloped to a new plan, while others were rebuilt, with some of the ruins restored. It is estimated that in England some 20–30 buildings damaged in the Blitz, and otherwise during the Second World War, now stand as ruins. Significance can be demonstrated for all of these, first in terms of the rarity of these structures, amounting on average to less than one bomb site per major targeted city (a few more survive in London), and second in the degree to which they represent a significant episode of twentieth-century world history. It is these ruins, and not gun emplacements and airfields, which serve to reify *civilian* casualties and *civic* destruction during the Second World War. They therefore have a significant role in: the commemoration of the War and of its casualties; providing a focus for educational initiatives which can be highly charged and engaging; highlighting the character and effects of 'Total War' and its impact on infrastructure and the population as a whole; and contributing to local character.

Second is the argument that regeneration of urban areas damaged by bombing is a better and more fitting memorial than hanging on to the past. This was the subject of debate surrounding the Hermitage site (east London) where a levelled area undeveloped since its destruction in the Blitz was proposed as the site for a major Thames-side redevelopment, albeit with memorial gardens (Ramsey 1997). As was said at a meeting of survivors at Hiroshima in 1951: 'If Hiroshima rises from the ruins left by the atom bomb to become a finer and more beautiful city than any other, won't that be a great thing for the world peace movement?' (Toyofumi 1994: 54). This argument had strength when rebuilding itself was a symbol of determination to make a new start whilst defying the bombing. However, considering Hiroshima's status now as International Peace City, it is the one bomb site to have been retained – the Atomic Bomb Dome – which provides the visual link to past events, and which acts as the focus for all commemoration and much tourist activity.

It can be argued that to retain these tangible reminders does not necessarily require conservation of entire structures, and that a facade may be sufficient for purposes of representation. But that denies the significance of space. St. Mary Aldermanbury in London, for example, bombed in the Blitz, stood as a ruined shell until 1965 when it was dismantled stone by stone and transported for re-erection in Missouri as a memorial to Churchill and his Iron Curtain speech there in 1946. Importantly, however, the site remains as open space, with the 12 Corinthian columns and part of the lower courses still in situ, giving the site a character in keeping with both its original function, the effects of war on civic pride, urban fabric and the wartime and post-war history of the building itself. It still therefore represents a space within an urban setting, where the effects of the Blitz can be recalled or imagined. While the ruin may contribute most to local character, the space it contains reflects better the impact of bombing, allowing the site to function as a heritage resource, a 'memorial park', as well as having value for other (for example nature conservation and quality of life) interests. Collectively these small enclosed spaces are an important feature of the urban scene, and are sustainable by virtue of being in scale with their close-knit townscapes.

Control Towers

In recalling the air war, control towers (or watch offices as they are sometimes known) provide a focus for the attentions of enthusiasts, historians, film makers and veterans. These are the structures which arguably best reflect the character of the air war, and which symbolise the losses incurred. It was from here that aircraft movements were controlled, and therefore it was here that losses were often first

Fig. 10 The control tower at former RAF Tangmere (West Sussex): an iconic building on the former Second World War fighter station. *Photo*: author

registered. While Operations Rooms have a similarly symbolic role in recalling these events, the control towers have a visual appeal which makes them obvious and iconic structures in the modern landscape (Fig. 10).

English Heritage's approach to managing control towers takes this into account. In all, some 450 control towers existed in England during the Second World War, some 220 of which survive in some form today. This c. 50% survival rate is much higher than for most other classes of Second World War monument and is likely to reflect the commemorative values these structures imbue and have imbued since 1945. By selecting the best preserved structures, and those with original fittings and fixtures, only those towers which have remained in use or which have adapted to new uses would be retained. Many of the towers are now ruined, however, and some of these no longer bear any obvious relationship to their flying field, or other components of the airfield such as hangars. Yet in the case of Battle of Britain airfields, or those associated with the bombing campaign or the Battle of the Atlantic, ruinous or isolated towers still remain significant for particular historic reasons, notwithstanding the social significance that is arguably a feature of all examples, however fragmentary they may be.

To balance these various factors, and to ensure any selection includes good surviving examples as well as those ruins whose values are more symbolic and connected with remembrance and commemoration, a set of criteria was developed for selecting control towers for protection in England. In short, the approach allowed control towers to be identified as significant for any, or a combination, of the following four reasons. First, where a well-preserved structure survives with original features, being exemplary of its type – and there were 18 main types of control tower in all (Paul Francis: personal communication): for example Duxford, which now houses part of the Imperial War Museum and has an unmodified tower, acknowledged as the best of its type. Second, where the tower stands on a site that has operational significance, such as Tangmere, with its strong ties with the Battle of Britain (Fig. 10). Third, where the site has historic interest, but for non-operational reasons: for instance Twinwood Farm, the station from which Glenn Miller flew on his ill-fated flight. Fourth, where the control tower has an obvious and visual relationship to contemporary surviving fabric or the flying field, such as Catterick, where the relationship to the grass airfield and airfield defences can be clearly seen.

By selecting sites on this basis, some 55–60 control towers would be identified as worthy of protection. Some have continued in use and remain as control towers; others are ruinous structures. Either way, a selection will remain – to remind future generations of the air war, a characterising feature of the twentieth century; to contribute to local character; to act as the focus for remembrance and commemorative events; and, at a more personal level, to serve as a catalyst for memory and remembering amongst veterans and their families.

Conclusions

This chapter has described the approach taken in England towards managing two specific categories of wartime remains, and the motivations for preserving such a selection of sites. It has also shown how those surviving monuments provide a focus both for commemorative events, and for remembrance as well as having historic interest. Often these are places for quiet reflection, but which also have strong visual impact, providing a physical record of significant wartime events.

The Second World War is remembered in many different ways in England: anniversaries, commemoration and remembrance services, visits to museums, in educational curricula, air shows, television broadcasts and other popular media. But what all of these also need are the physical, tangible places where these events unfolded. These are not just historic sites, like prehistoric burial mounds and hill forts; these are also memorials to the events of the Second World War, and to recent conflict generally. As with the physical remains of First World War battlefields, for example at Vimy Ridge and Beaumont Hamel (Cave 2000), these are often sacred sites and as such should engage the visitor, evoking some response which brings the events of 60 years ago into sharp focus. Arguably, those sites associated directly with loss of life, personal tragedy and civic destruction to cultural property provide the best opportunities for engaging the events of the Second World War in this way. It is primarily for that reason that representative examples of these classes of monuments are being protected or retained in England.

Section 2
Memory and Place

Places are powerful things; emotionally draining, demanding, sometimes joyous, and familiar or (at times) unfamiliar to those that inhabit or visit them. We feel close to places, comforted by their existence, by the fact that sometimes they don't change – Gidden's ontological security (1991). Sometimes we are encouraged by the fact that they do – that places can be given a new lease of life, or new meaning and significance. But places can also be challenging and difficult, often because they are contested. People from different social or cultural backgrounds might value the same place in different ways for different reasons, and those reasons can be contradictory, especially where militarised landscapes are concerned. At Greenham Common for example, as Chap. 7 reveals, some people value the military heritage of the former airbase for what was achieved there, for its contribution to ending the Cold War maybe; some value the peace camps and the material culture of protest and opposition (perhaps for exactly the same reason!) and some value the business park that has developed on the former technical site, for the jobs and opportunities created, giving something back after the military intervention. Then there is the Common, now publicly accessible after decades of closure and secrecy, and the commemorative park and sculptures raised by the peace women whose caravan occupied the site in the 1980s and 1990s. Finally, the fence, dividing one set of cultures and values from another. The fence still stands in places: one of the front lines of the Cold War. One place – many values, and many statements of meaning and significance.

But for this section we take a step back, and think what we mean by place and how this fundamental construction relates specifically to conflict archaeology. To really understand 'place', one can usefully begin with some of the theoretical principles given clearest expression in the field of human geography. Tim Cresswell for example sees place as a way of,

> Seeing, knowing and understanding the world. When we look at the world as a world of places we see different things. We see attachments and connections between people and place. We see worlds of meaning and experience. Sometimes this way of seeing can seem to be an act of resistance against a rationalization of the world, a way of seeing that has more space than place. To think of an area of the world as a rich and complicated interplay of people and the environment – as a place – is to free us from thinking of it as facts and figures. To think of Baghdad as a place is in a different world to thinking of it as a location

on which to drop bombs. At other times, however, seeing the world through the lens of place leads to reactionary and exclusionary xenophobia, racism and bigotry. 'Our place' is threatened and others have to be excluded. Here 'place' is not so much a quality of thing in the world but an aspect of the way we choose to think about it – what we decide to emphasise and what we decide to designate as unimportant. (2004: 11)

The places I describe in this section are places I have deliberately chosen to study, for reasons that I will try to explain. But to summarise and probably oversimplify it, these are places which are all contested, which will continue to be so, and which are fascinating for the various ways in which this contest continues to be played out in the academe, heritage practice and amongst local communities. There is exclusion here, there is reaction and there is resistance. But above all, we see connections and attachments between people and place, and place and memory.

In Chap. 4, my subject is Berlin. It will be obvious on reading the chapter, but here I collaborated posthumously with my father who was stationed in Berlin in 1971–1973. It is a time I remember well and recall fondly (see also the Afterword). I attended a symposium at Cecilienhof Palace (Potsdam) in 2004 to discuss preserving monuments and sites of the Cold War era. I co-presented an overview paper on English Heritage's study of Cold War architecture and archaeology. But the symposium also included visits to some of the sites we were discussing, including the Allies' museum in the former American Sector, where I talked about my experiences, as a child in Berlin at the height of the Cold War. I was invited to contribute an essay on this subject to the conference proceedings and it was while researching that essay that I discovered my father's lecture notes, for a series of talks about Berlin that he gave in his retirement to local groups in Suffolk (England). By combining his notes and my memories the essay came together.

My memories here are of what mostly appeared to me then as a mundane and an ordinary place, characterised by everyday journeys to school and after school clubs. But there were extraordinary moments that are vivid memories for me. Such as taking a late night tour of the Wall, and seeing East German border guards scanning us with their binoculars. The Wall dominated our lives, even though we did not often see it – it had a presence; it performed us, as Bourriaud (2002) would say. Leo Schmidt's Introduction to the conference proceedings describes perfectly the relevance of place in relation to memory, to remembering and to not forgetting:

Memory clings to places and objects. Objects, and buildings in particular, are identified with memory. By consequence, many intact buildings all over the world have been destroyed because they stood for a painful memory whilst other buildings, whose destruction by war or catastrophe could not be tolerated, have either been recreated or are the focus of highly emotional debates on reconstruction. The Berlin border provides examples for both positions, but so far the wish to extinguish history by extinguishing its witness has been predominant. (2005: 16).

By removing the object that gives our memories a focus, a centre-point, the memories of everyday events and actions can also be lost. At least with the Wall the void has become the monument – the vacuum its most monumental remnant.

Chapter 5 presents another contested place, where remembering through material culture and monumental architecture occurs on different levels: particular nuclear protest events at Peace Camp that generated stone circles or artefact deposition; scientific research programmes within the Nevada Test Site; and the concerns and claims of traditional owners over the wider landscape. There are also deeper more embedded memories within this landscape and attempts at erasure in the name of neo-colonialism and for 'the good of all'. And there are more recent, transient interactions from those just passing through. As with the Berlin Wall there is a dividing line here, one that still exists, enclosing the Test Site and all that lies within it. Indeed, divisions exist within the Test Site too, dividing the extent of particular research programmes, or the domain of individual research laboratories, including famously that containing Area 51 – one of the most secret places on earth.

To what extent the legitimate archaeology of the Test Site, and funding through the US Department of Energy, contrasts with the alternative and – some say – subversive archaeology of Peace Camp was my motivation here. I had read about the Test Site some years before, and a simple magazine article (Johnson and Beck 1995) inspired my interest in this other worldly and remote desert landscape. But surely with so much atomic testing taking place, and the growing concerns for health, not to mention global meltdown, there had to be some opposition to all of this? And, in such a remote place, the material culture of that opposition should remain. We were told it did not, and that even if it did, it was not the concern of archaeologists. That was enough for me. With Colleen Beck and Harold Drollinger of the Desert Research Institute (Las Vegas), and with no funding to speak of, we conducted two seasons of work at Peace Camp. The results of that study are outlined in Chap. 5.

Twyford Down has particular memories for me. It lies close to the university where I completed my undergraduate and postgraduate degrees, and I walked there frequently from my rented house in Winchester, dodging the traffic on the old, narrow A33 to reach the foot of the hill. I recall my first journey to university, with my parents sitting proud in the front of the car, all my records and record player safely boxed in the back, and a stationery queue of traffic as far as the eye could see. The decision to place a new road through Twyford Down, thus removing the undisputed problem of the old 'Winchester bypass', was hugely controversial, and a defining moment in the environmental protest movement in Britain. My formal involvement in this began as English Heritage set about revising the boundaries of the protected scheduled monument in the area following construction of the new road, and my determination to include within that protection some evidence of the recent controversy, the latest of many changes to this landscape, all made in the name of progress and transportation.

To cite the original abstract, Chap. 6 goes to the heart of many of the accepted notions that inform heritage practice and theory: of the permanence of monuments; their legitimisation by age; their preservation from change and their representation of a social consensus. By contrast, modern 'intrusions' to lived space are designed to be impermanent, are obviously new, represent change and often result from conflict. Twyford Down is an example – a concrete expression – of this discordance: it has legal protection, but was compromised by the construction of the M3 motorway extension in the late 1980s. Yet, with archaeologists increasingly willing to explore the contemporary past, can sites like Twyford Down not be interpreted in a very different way, by recognising the landscape as dynamic not static, and by understanding that the process of change is as relevant today as it was in the past? In this essay, such an interpretation of landscape and heritage management practice is suggested, placing Twyford Down's later twentieth-century components alongside those of earlier date.

That then brings us a short distance north – about 20 min by car – to Greenham Common, the former airbase that hit headlines in the 1980s and 1990s with the arrival of cruise missiles and a mass protest. The Greenham Women embracing the base has become a defining image of the twentieth century. Like many of the essays in this collection, this one originated in a chance encounter. I had been to a site visit on the airbase, at around the time of its drawdown and closure. We finished early and were driving into town to drop people at the railway station. One of those in the car recognised something, and urged me to turn right up a very narrow wooded lane off the main road. He recognised this place as leading to the camp his wife had attended some 15 years earlier. Being a man he had not been to the camp, but he knew where it was, and he had seen the access road often enough. So we drove to the end to face the stunning sight of vast missile shelters in close up, with only a tattered fence separating us from them. We wondered around in the trees and found what we thought must be remnants of the campsite.

This brief encounter led to a long fascination with Greenham, and ultimately my work in Nevada. In both cases, I was interested in the contradictions, in material remains, their size, scale, their ephemerality, transience and permanence, and in official- and locally held attitudes and opinions. Why was it that the military estate had popular support and cultural accreditation, while the evidence for protest and opposition (which after all is just as 'archaeological') was ridiculed and dismissed as 'rubbish' (in at least two senses)?

This Greenham article is one of the first that I wrote on the subject of recent archaeology, appropriately in a volume on the subject of queer archaeologies which, as the editor explained, 'challenge all aspects of established normative practice'. He notes how the essay provides an

> atypical, outlandish archaeology of a Cold War site in England. … In so doing [this] archaeology challenges established methodologies for twentieth-century archaeology and heritage management. It presents a controversial theme as legitimate archaeology, controversial because it doesn't fit established, normative practices in archaeological and scientific discourse. It challenges the epistemological privilege inherent in archaeology in particular and science in general. (Dowson 2000: 164)

Finally, in this section, a summary of my most recent project, exploring a single street in the World Heritage Site city of Valletta (Malta) is described. Not much conflict there you might think, but you would be very wrong. This is a place with such stigma attached, so much taboo, where so little is said, that the street has been abandoned and left empty and unloved for about 40 years. People rarely go there, especially visitors but locals too. That degree of closure and denial fascinates me, as do (and this should be obvious by now) places which have hidden histories, mysteries and myths.

Chapter 8 was co-written for a themed issue of the Journal of Mediterranean Studies. This particular issue represented the culmination of a European Commission funded EURO-Med Heritage project on Mediterranean Voices. Valletta was one of the partner cities in this project and (here is another coincidental encounter) we were researching Strait Street at the same time the Valletta branch of this project was exploring and documenting the city's hidden places. The connections were so strong between our projects that we were invited to contribute a summary of our research to the volume: 'Mediterranean Voices: Turning back to the Mediterranean'. It is that which is repeated here.

The editors of the volume describe our contribution as representing a view of imagined or remembered space. Strait Street was (and remains) a street of shame for many but it was also the centre of a vibrant entertainment district. They draw comparison between Strait Street and the 'street of love' in Nicosia and the events of the Barrio district of Ciutat (Bartolomé and Tipper 2005). As Radmilli and Selwyn say,

> Strait Street was, and is, not alone in being the 'gut' of a city. Recalling that the Mediterranean outpost of London's Soho was the space inhabited by the first Italian immigrants to London, as well as being famous for bars, clubs, restaurants and brothels, it seems reasonable enough to suggest that, if it is cosmopolitan co-existence, cultural creativity, and general urban well-being that you want, city planners need to listen to a wide range of voices including those who speak, as it were, directly from the 'gut'. (2005: 207)

The chapters included here are reproduced by kind permission of the original publishers. Original citation details are as follows: Schofield J. and Schofield Gp Cpt A. (2005), Views of the Wall: Allied Perspectives. In Schmidt, L. and von Preuschen, H. (eds), *On Both Sides of the Wall: Preserving Monuments and Sites of the Cold War Era*, pp. 36–43. Berlin: Westkreuz-Verlag; Beck, C., Drollinger, H. and Schofield, J. (2007), Archaeology of dissent: Landscape and symbolism at the Nevada Peace Camp. In Schofield, J. and Cocroft, W. (eds), *A Fearsome Heritage: Diverse Legacies of the Cold War*, pp. 297–320. Walnut Creek: Left Coast Press One World Archaeology 50; Schofield, J. (2005), Discordant Landscapes: Managing Modern Heritage at Twyford Down, Hampshire (England). *International Journal of Heritage Studies* 11.2: 143–159; Schofield, J. and Anderton, M. (2000), The queer archaeology of Green Gate: interpreting contested space at Greenham Common Airbase. *World Archaeology* 32.2: 236–251; Schofield, J. and Morrissey, E. (2005), Changing Places – Archaeology and Heritage in Strait Street (Valletta, Malta). *Journal of Mediterranean Studies* 15.2: 481–495.

Chapter 4
Views of the Berlin Wall: Allied Perspectives

With Gp Cpt Arthur Schofield posthum

This chapter provides two related perspectives on the Allied experience of Cold War Berlin: one is my own, that of a 9–11-year-old child living with my parents within the confines of RAF (Royal Air Force) Gatow from 1971–73; and the other is that of my father Arthur Schofield, then a Wing Commander in the RAF. The chapter includes extracts of both our perspectives with our names to identify each. My notes are just that: notes of memories prompted in many cases by photographs, a couple of which are included here. They are selective, and quite clearly childhood memories that are included, recalling the excitement of Berlin at this time. My father's account is published posthumously, and taken virtually verbatim from lectures he gave to local societies and groups in Suffolk following his retirement from the RAF in 1977. His notes were written in about 1990. He provides general descriptions and information that will be familiar to some readers, and historical context to what I have written. He did not (and perhaps could not) reveal details of his own role in Berlin, though it can now be told that he was Officer Commanding 26 Signals Unit (Fig. 11), based at the listening post in the Grunewald known as Teufelsburg, or 'The Hill' (see Haysom 2002 for some background on Signals Units in Berlin). I include no summary or analysis with this contribution, merely the accounts themselves, providing a personal note to the historical phase of Cold War Berlin.

I should add that, towards the end of a long service career, Berlin was the posting my parents most wanted. Not only was it a challenging and exciting post for those involved with signals and intelligence, but the social and cultural diversity, coupled with it being the Cold War's front-line, made it the most desirable move for many. Also Berlin is, and was then, a wonderful city. In fact, not realising we were to be posted there, we spent the previous Christmas on holiday in Berlin, ensuring we visited the city before our expected move from our current posting in Germany back to the UK.

Perspectives

AS: The Russians captured Berlin on 2 May 1945 and then swept on, with Allied Powers' agreement, for another 104 miles to meet up with UK and United States forces on what was to become the boundary between West and East Germany.

J. Schofield, *Aftermath: Readings in the Archaeology of Recent Conflict*,
DOI: 10.1007/978-0-387-88521-6_5, © Springer Science+Business Media, LLC 2009

Fig. 11 My father (*front row, centre*) and others of 26 Signals Unit, Berlin c. 1972. *Photographer* unknown

Berlin, with 2.4 million citizens, was declared a city under four-power military government and divided into four sectors: American, British, French and Russian. Each sector was under the command of a Military Governor supported by Protective Armed Forces, a four-power Air Traffic Control Agency, and civilian diplomatic and administrative staffs. The German population of the city was under military command and control with representatives forming the Berlin Senate which paid all the Allied occupation costs.

In practice, Berlin was divided into two areas, the Americans, British and French forming the Western area (some 65% of Berlin), and the Russians the Eastern (35%). Access to each of the western sectors was by prescribed and controlled air, road and rail corridors, with entry points at the West German border, and at the Berlin end. For the British, the RAF Station at Gatow provided the air link, and army battalions were stationed at the Spandau and Gatow barracks near the airfield, in order to defend it.

The Allies set about restoring their own sectors, their economic recovery gradually making West Berlin an attractive magnet for East Berliners – little had been done in East Berlin to repair war damage and restore industry and commerce. The result was an increasing flow of East Germans into the western sectors, despite attempts to prevent the loss of skilled workers. Matters came to a head in 1961. At two o'clock on the morning of 13 August, the western sectors were cut off by barbed wire, and barricades were erected by the People's Police and the National People's Army. Two days later, the building of a concrete wall one-and-a-half metres high was started, buildings close to the Wall were demolished, and despite protests from the Western governments, a wall 12 km long and a fence of barbed wire some 137 km long were completed to make West Berlin an island, some 190 km within East Germany.

JS: While stationed at Gatow, my father often had visitors from England (I used to tell school friends we had VIPs staying with us). Some I now know were from GCHQ – signals and intelligence staff I presume, though I didn't know this at the time. My parents used to entertain them, typically including a restaurant dinner in Central (West) Berlin. When I was slightly older I was allowed to go too. After a pizza, we travelled to the Wall and looked over into the East, late at night, from some of the viewing platforms that existed then. The guards trained their binoculars on us, and the dogs in the Death Strip barked. Otherwise it seemed all the noise, of a city continuing to live through the night, came from the West; the East was silent and eerie, even more so than during the day. Once some friends stayed the night at our house on RAF Gatow, some five kilometres from the border with East Germany. One morning my mother took a tray of tea into their bedroom; they were sitting upright in bed and terrified. They had been woken at about four o'clock in the morning by the sound of machine gun fire, which had sounded close as the wind was from the west. We were used to it. Machine gun fire always sounded when there were escape attempts.

AS: Before the Wall, access between the sectors for civilians and military personnel was in theory not restrained, despite Russian and East German police provocation at times. With the building of the Wall, communications were disrupted and West Berliners were completely cut off from relatives and friends in the Eastern Sector. On 17 August, the first East Berliner trying to cross the Wall was shot, and by February 1989, it was estimated that some 250 deaths had occurred at the Wall or barbed wire barricades.

Checkpoint Charlie was the only entry point for US and UK personnel into the Russian sector, while Checkpoints Alpha and Bravo were the entry points at the West Germany and Berlin ends of the linking Helmstedt Autobahn. All three checkpoints were manned by UK Military Police and Russian or East German Military Police, on either side of the border. All movements had to be covered by documentation, which had to be carefully checked, stamped and recorded (Fig. 12).

Movement between the US, French and British sectors was unrestricted, and recreational and social activities for service personnel were not affected.

JS: My mother took me into the East at least twice, in a British staff car with a military police escort. We passed from the West through Checkpoint Charlie, amidst all the usual security checks. They looked at our passports and Movement Orders and then our faces through the car's closed windows. I stood on the nearly deserted Unter den Linden, just beyond the Brandenburg Gate – in another world – and had my photograph taken. Separately – on an organised coach tour I think – we visited the Soviet War Memorial at Treptow and I was too frightened to leave my mother and visit the men's toilets. A stern-looking middle-aged woman was cleaning out the ladies'. My mother confidently approached her and asked in her broken German if we could both go in together. The cleaner looked at me and must have taken pity. Her stern expression was replaced by a warm smile as she ushered us both in.

Fig. 12 Movement Order, giving passage for my mother and me on the Berlin-Helmstedt Autobahn, between 5–8 March 1973 (Author's collection)

While my mother and I were together in the East, my feelings were of immense excitement, though my mother I suspect was apprehensive. However, once there was a danger of us becoming separated, I became concerned, as at Treptow and at the large museums where I feared us just losing each other in the cavernous spaces and labyrinthine galleries. I remember at least once having nightmares about this. My father was strictly forbidden from going to the East, because of the dangers of capture and interrogation. My mother has since told me that she was briefed about what not to say if captured. Needless to say, I never had any such guidance. Interestingly, the ability to travel into East Berlin was not a privilege extended to senior American servicemen or their families; yet only very few British military personnel, and none of their families, were prevented – or discouraged – from travelling to the East.

The excursions I recall involved our British military police escort being followed, and very obviously so. I remember a car trailing us everywhere, with two men in long coats and dark hats. I remember upturned collars but I'm not certain how far that is memory or merely the influence of the spy films I have subsequently seen. However, one time we stopped to buy some porcelain. My mother and I went into the shop with one of our escorts, the other remaining in the car. The East German car pulled up behind ours, and the two men came to the shop window. I clearly remember looking out the window at the two men looking in. Our eyes met. I thought of them as the enemy, and didn't smile. They looked back at me, straight-faced. I felt no fear, just a kind of exhilaration.

As well as porcelain, I also remember my mother buying a Russian hat for my father, and one for my older brother, also in the RAF at that time. My mother also bought one for me, though I hadn't noticed that. My brother came to stay for Christmas and we were all given our hats. We all then stood for a photograph, in the garden of our house (Fig. 13).

Fig. 13 My father, me and my brother (*left–right*) in our garden in Berlin, Christmas Day 1972, with our new Russian hats. *Photo*: author's collection

Being a virtual island, every holiday involved a trip down the 'corridor' between Berlin and Helmstedt, and participating in all the rituals that this inevitably involved. There were clear and strict guidelines on how to behave and on appearance, both at the border crossings and within East Germany. I had to be both well behaved and smartly dressed. In the 'Regulations Governing Travel on the Berlin-Helmstedt Autobahn' it states that:

...

6. The GOC [General Officer Commanding] Berlin (British Section) has decreed that all motorists travelling to and from Berlin on the Helmstedt Autobahn under military sponsorship are to maintain a high standard of appearance in their dress and the highest standard of deportment in their contact with the Russians at the controlled checkpoints en-route.
7. The GOC, like any other traveller down the corridor, has been much impressed with the courtesies extended and the manner of liaison with the Russian military staff encountered on the journey and it is only common courtesy that British personnel dress and act in a manner befitting their own interests and those of the interest which they represent.
8. In future, therefore, travellers on reporting to [the] checkpoints at Grabert Bridge, Dreilinden and Helmstedt are to be suitably attired, if travelling in civilian clothes, in shirt with collar and tie, long trousers, shoes and socks. Anyone not conforming to these standards will not be allowed to proceed.

During my time in Berlin, we occasionally used the British military train that ran from West Berlin to Braunschweig (Fig. 14).

We also once took the French military train to Strasbourg. This was an overnight train, manned by the French, and guarded by French military personnel. All the usual codes of dress and behaviour applied. I remember being irritated that taking this trip with my parents took me out of school, and an arranged visit to Berlin Zoo. Now a carriage of this French military train stands at the Allied Museum (Alliierten-Museum Berlin) on Clayallee.

When we left Berlin in 1973 my mother and I flew to England. My father drove with the car and caravan, the caravan filled with some of our possessions. As always, he had a police escort for the 2-h drive from Berlin, through East to West Germany. We knew he was leaving Berlin at dawn, and due in England late that evening. But he didn't turn up when we expected him to, and we had heard nothing from him. My mother and I were staying with my sister. I recall in the middle of the night a knock on the front door. I was in the bedroom above and heard my mother hurry downstairs. A policeman was at the door and conveyed a short message, which was all he had received, to tell my mother that my father had had an accident in East Germany. I was scared, and feared the worst, though wasn't sure what that was. Early next morning though a more detailed message came through, that the caravan had broken an axle on the potholes in the autobahn, and that – having a car and caravan – the only means of rescue was a tank transporter. So he had sat all day, playing cards with his armed police escort and reading Agatha Christie books from the caravan, waiting for the transporter to arrive. He had then been taken to West Germany. He refused to return to Berlin after so many had turned up at dawn to wave him off.

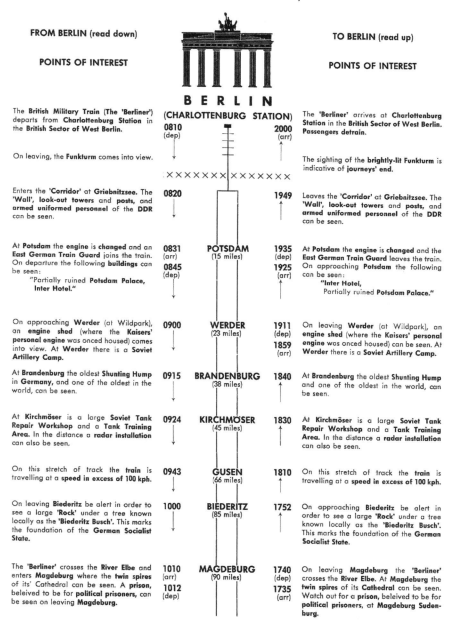

BRITISH MILITARY TRAIN
(BERLIN — BRUNSWICK — BERLIN)

FROM BERLIN (read down)

POINTS OF INTEREST

TO BERLIN (read up)

POINTS OF INTEREST

B E R L I N
(CHARLOTTENBURG STATION)

FROM BERLIN			TO BERLIN
The **British Military Train** (The '**Berliner**') departs from **Charlottenburg Station** in the **British Sector of West Berlin.**	0810 (dep)	2000 (arr)	The '**Berliner**' arrives at **Charlottenburg Station** in the **British Sector of West Berlin. Passengers detrain.**
On leaving, the **Funkturm** comes into view.			The sighting of the **brightly-lit Funkturm** is indicative of **journeys' end.**
Enters the '**Corridor**' at **Griebnitzsee.** The '**Wall**', **look-out towers** and **posts**, and **armed uniformed personnel** of the **DDR** can be seen.	0820	1949	Leaves the '**Corridor**' at **Griebnitzsee.** The '**Wall**', **look-out towers** and **posts**, and **armed uniformed personnel** of the **DDR** can be seen.
At **Potsdam** the engine is **changed** and an **East German Train Guard** joins the train. On departure the following **buildings** can be seen: "Partially ruined **Potsdam Palace, Inter Hotel.**"	0831 (arr) 0845 (dep)	1935 (dep) 1925 (arr)	At **Potsdam** the **engine** is **changed** and the **East German Train Guard** leaves the train. On approaching **Potsdam** the following can be seen: "**Inter Hotel**, Partially ruined **Potsdam Palace.**"
On approaching **Werder** (at Wildpark), an **engine shed** (where the **Kaisers' personal engine** was onced housed) comes into view. At **Werder** there is a **Soviet Artillery Camp.**	0900	1911 (dep) 1859 (arr)	On leaving **Werder** (at Wildpark), an **engine shed** (where the **Kaisers' personal engine** was onced housed) can be seen. At **Werder** there is a **Soviet Artillery Camp.**
At **Brandenburg** the oldest **Shunting Hump** in **Germany**, and one of the oldest in the world, can be seen.	0915	1840	At **Brandenburg** the oldest **Shunting Hump** and one of the oldest in the world, can be seen.
At **Kirchmöser** is a large **Soviet Tank Repair Workshop** and a **Tank Training Area.** In the distance a **radar installation** can also be seen.	0924	1830	At **Kirchmöser** is a large **Soviet Tank Repair Workshop** and a **Tank Training Area.** In the distance a **radar installation** can also be seen.
On this stretch of track the **train** is travelling at a **speed in excess of 100 kph.**	0943	1810	On this stretch of track the **train** is travelling at a **speed in excess of 100 kph.**
On leaving **Biederitz** be alert in order to see a large **'Rock'** under a tree known locally as the '**Biederitz Busch**'. This marks the foundation of the **German Socialist State.**	1000	1752	On approaching **Biederitz** be alert in order to see a large **'Rock'** under a tree known locally as the '**Biederitz Busch**'. This marks the foundation of the **German Socialist State.**
The '**Berliner**' crosses the **River Elbe** and enters **Magdeburg** where the **twin spires** of its' Cathedral can be seen. A **prison**, believed to be for **political prisoners**, can be seen on leaving **Magdeburg.**	1010 (arr) 1012 (dep)	1740 (dep) 1735 (arr)	On leaving **Magdeburg** the '**Berliner**' crosses the **River Elbe.** At **Magdeburg** the **twin spires** of its Cathedral can be seen. Watch out for a **prison**, believed to be for **political prisoners**, at **Magdeburg Sudenburg.**

Station names (centre column): POTSDAM (15 miles), WERDER (23 miles), BRANDENBURG (38 miles), KIRCHMÖSER (45 miles), GUSEN (66 miles), BIEDERITZ (85 miles), MAGDEBURG (90 miles)

Fig. 14 The 'Guide', given to passengers on the British military train (Author's collection)

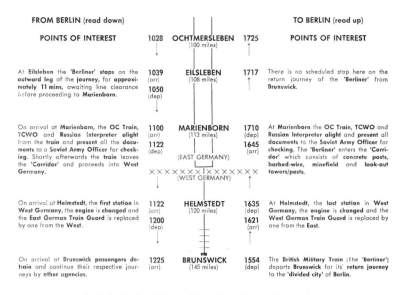

FROM BERLIN (read down)					TO BERLIN (read up)
POINTS OF INTEREST	1028	OCHTMERSLEBEN (100 miles)	1725		POINTS OF INTEREST
At **Eilsleben** the '**Berliner**' stops on the outward leg of the **journey**, for approximately **11 mins**, awaiting line clearance before proceeding to **Marienborn**.	1039 (arr) 1050 (dep)	EILSLEBEN (108 miles)	1717		There is no scheduled stop here on the return journey of the '**Berliner**' from **Brunswick**.
On arrival at **Marienborn**, the OC Train, TCWO and **Russian Interpreter** alight from the **train** and **present** all the **documents** to a **Soviet Army Officer** for checking. Shortly afterwards the **train** leaves the '**Corridor**' and proceeds into **West Germany**.	1100 (arr) 1122 (dep)	MARIENBORN (113 miles) (EAST GERMANY)	1710 (dep) 1645 (arr)		At **Marienborn** the OC Train, TCWO and **Russian Interpreter** alight and **present** all **documents** to the **Soviet Army Officer** for checking. The '**Berliner**' enters the '**Corridor**' which consists of **concrete posts**, **barbed-wire**, **minefield** and **look-out towers/posts**.
		× × × × × × × × × × × × × × × (WEST GERMANY)			
On arrival at **Helmstedt**, the first station in West Germany, the **engine** is **changed** and the East German Train Guard is replaced by one from the **West**.	1122 (arr) 1200 (dep)	HELMSTEDT (120 miles)	1635 (dep) 1621 (arr)		At **Helmstedt**, the **last station** in West Germany, the **engine** is **changed** and the West German Train Guard is replaced by one from the **East**.
On arrival at **Brunswick** passengers detrain and continue their respective journeys by **other agencies**.	1225 (arr)	BRUNSWICK (145 miles)	1554 (dep)		The **British Military Train** (the '**Berliner**') departs **Brunswick** for its' **return journey** to the '**divided city**' of Berlin.

The '**Berliner**' is shunted into the sidings at **Brunswick** where it is cleaned and prepared for its' return journey to the '**divided city**' of **Berlin** on the same day.

ROUTE OF BRITISH MILITARY TRAIN
(THE 'BERLINER')

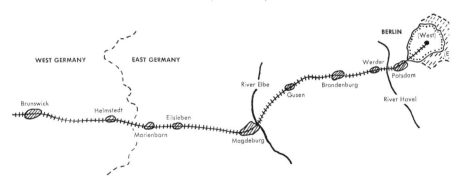

Fig. 14 (continued)

There was much cultural diversity in Cold War Berlin, which I am pleased to have experienced. This diversity manifested itself in the different nationalities present within the city, but not within the service environment of which we were a part. Socially for example, officers' families rarely mixed with those of other ranks, and similarly there was little contact – for the children certainly – between Army and Royal Air Force families. I recently revisited Berlin with my children and took them to swim at the outdoor Olympic swimming pool, once within the British sector, and constructed for the 1936 Olympic Games. They loved it, and I asked my mother if I had ever been there. 'Oh no, dear – that's where the Army families swam. We had the [RAF] Officer's Club!'

But children of my age did mix at school, and my particular friends were from army backgrounds and other ranks in the Army at that (probably those from the Gatow barracks, charged with guarding the airfield). I still remember most of their names. But it was an odd friendship because we never met outside of school. One weekend I invited my two best friends around to my house to play (my parents had agreed, I suspect reluctantly). But as it turned out, their parents were even less happy about their boys visiting the son of the man running 26 SU, and they didn't turn up. That's how it was. Social class reflected by service rank was rarely overcome. Out of school I tended to play with the girls that lived in the flat upstairs and the house next door (of which more in the Afterword); In school it was the Army boys – Two separate worlds; the story of the city.

One of the advantages of the Berlin posting was the contact it encouraged with French, Americans and Germans. The social life for my parents was rich and diverse, and the babysitter I was allocated became almost like an older sister. For me it was a wonderful education. On Sundays we would sometimes travel to the other sectors for lunch; the French Maison des Cadres for a lunch of many courses, where I had escargot for the first time, and natural yoghurt, and sat seemingly for hours while others talked. I remember the cobbled streets and architecture of this part of the city, which appeared to me 'very French'. I had picked up on the feel of the place, its essence, its personality, being French: road names and voices, and the smell of French cuisine. And then the American sector, where we used to shop at the 'PX', and buy T-shirts, and attend picnics hosted by the American servicemen at their camp. It was at one of those picnics that I had my first beef burger, and I remember noticing the very different personality, attitude and behaviour of the American servicemen. They seemed to joke more than we did, their sense of humour was less subtle than our own, and they drank from bottles. Part of this camp is now the Allied Museum.

At a function at the Officer's Mess, my parents were introduced to an Italian couple – the wife was an opera singer. They lived in a flat somewhere in the city where we once went for dinner. She breast-fed her baby at the dinner table and there were huge cages of birds – green- and goldfinches mainly – in some of the living rooms. The Italian couple were both very expressive. I loved the dark corridors and tiled floors of their flat, and the birds caged high above the city streets, behind their own 'Berlin Walls'. We also had some German friends, including a retired Naval commander. Once at dinner at his villa on the banks of the River Havel he presented me with a box of miniature lead German ships; one of each class in the wartime German Navy. I still have them, of course.

AS: Throughout 1989, industrial unrest in the Eastern Sector and East Germany increased, with pressure to bring about unification. In November, the crossing points were thrown open and controls lifted. West Berlin was overwhelmed from the East by half a million visitors, all receiving a handout of West German marks as 'welcome money'. Pieces of the Wall were broken up and sold as souvenirs. In December 1989, East and West German governments agreed to abolish border stations, and the Brandenburg Gate was opened to pedestrians on 22 December. The Wall was gradually broken up and pieces sold to provide funds for the East German health-care system.

After the Wall

My father never returned to Berlin. I have been back several times, the last for the conference at which this chapter was originally delivered. I climbed the hill to look at Teufelsburg through the high perimeter fence that surrounds it (the plans are now to convert it into a hotel and leisure complex). I have returned to Gatow and visited my house and school, and I visited the Allied Museum, where a carriage from the French military train is on show. I have also read books such as those by Garton Ash (1997) and Funder (2003) which discuss Stasi archives, and wonder what information they have about my father, if anything. Many of the contributions to the original volume, of which this chapter was a part, describe the physical and emotional legacies of the Wall. This contribution shares that objective, but from a personal rather than a cultural perspective. I make no judgement on the cultural significance of the buildings at Gatow and Teufelsburg, or of the military train – though its retention within a museum context may be significant. I value these places for my own reasons, and for the entrée they provide to my own experiences of Berlin. These places have social significance and values for these very intimate reasons, in addition to any cultural values that others may determine.

Returning to our regular journeys down the 'corridor', I was interested, at the time the Wall was being demolished, to hear the fate of the checkpoint at Helmstedt, and the motel – reserved only for service personnel – at which we regularly stayed. A *Daily Telegraph* newspaper article dated 4 October 1990 went some way to answering those questions: '… At nearby Helmstedt, there was not an official in sight as traffic in both directions on the motorway roared across the old border completely unhindered. The complex area of Customs posts, guardhouses and barbed wire and passport offices was like a ghost town. Little East German Trabants and West German Mercedes were parked in the once-forbidden no-man's-land as visiting families scampered up deserted East German observation towers'.

The change was remarkable in such a short space of time; and not just buildings and places, border installations and the Wall, but a whole way of life determined and driven by the political cultures of the Cold War. Yet the memories remain of an exciting time, in a divided land.

Chapter 5
Peace Camp, Nevada

With Colleen M. Beck and Harold Drollinger

Competing ideologies and the threat of nuclear war were central to the Cold War as the former Soviet Union and the United States engaged in a stalemate for military superiority (Halle 1967). The world lived under the spectre of a Doomsday Clock showing the minutes to midnight, the hour of nuclear war. Governments sought to protect their countries and citizenry through alliances and the development of increasingly sophisticated nuclear weaponry and delivery systems (Angelo and Buden 1985; Baker 1996).

These military efforts and the built environment associated with them are dominant in Cold War heritage. For example, the history of a United States nuclear weapons complex explains the roles of various, inter-related facilities in the design, development, production and testing of nuclear weapons (Loeber 2002), while on the other side of the Atlantic, a historic survey of the Atomic Weapons Research Establishment buildings and structures documents Great Britain's nuclear weapons development and design facility (Cocroft 2003). Others have focused on particular aspects of the nuclear weapons story, such as the Hanford plutonium production facilities (Marceau et al. 2003) and nuclear testing remains at the Nevada Test Site (Beck 2002). Even structures whose purpose was to study the Cold War sky through radar systems have been systematically recorded (Whorton 2002) and the civilian aspect has not been overlooked with the architectural designs of this era interpreted as reflecting the nuclear threat and Cold War politics (Johnson 2002). The publication of a broad overview of nuclear testing buildings and structures throughout the United States captures the nature of the Cold War era and shows that such topics have become mainstream (Vanderbilt 2002).

During the Cold War, however, there was some visible dissent with the dominant government actions and policies by pacifists and anti-nuclear activists. Some of the notable pacifists, such as Lillian Willoughby, Albert Bigelow and Ammon Hennacy, had protested against wars before the dawn of the nuclear age and their activism was only heightened by the emergence of nuclear weapons into the world's battlefields. Some others, in response to the devastation in Japan with its attendant visual impacts, protested the new weapon and its magnitude. Ultimately as the years passed, the belief that the Superpowers would eventually fall into war, annihilating populations on the earth, with survivors facing a nuclear winter, produced anti-nuclear activists throughout the world.

J. Schofield, *Aftermath: Readings in the Archaeology of Recent Conflict*,
DOI: 10.1007/978-0-387-88521-6_6, © Springer Science + Business Media, LLC 2009

One avenue for expressing alternative views was to conduct protests at sites iconic of the Cold War. These government facilities included missile silo arrays, air bases with bombers on 24-h alert, laboratories developing new nuclear weapons, plutonium-processing plants, a white train carrying nuclear materials across the United States, and nuclear-testing facilities. However, most protests in the United States and other countries have taken place in the paved world, on streets and in parks and parking lots, leaving little, if any, material remains of the protesters' activities. These centralised locations provide ease of access for the protesters and news media as well as being highly visible to people in the community. Following such actions is the inevitable clean-up and a return to normality of daily activities. An exception to this scenario is the situation where protesters established a semi-permanent camp just outside the entry to the Nevada Test Site. On a reduced scale, a similar situation also existed at Greenham Common Airbase in England where protesters' vigils also led them to camp at the location (see Chap. 7, this volume). The protests at the Test Site differ from most protest circumstances because they occur in a desert landscape, remote, and without facilities common in an urban setting. This chapter discusses the archaeo- logical study of this place, known as the Peace Camp. The research focuses on exploring the materiality of the occupation and the use of landscape and space in order to understand the nature of protest occupations.

Context

The Peace Camp is in southern Nevada adjacent to the Nevada Test Site, a limited access, government-controlled facility, covering approximately 3,600 sq km. The camp and the Nevada Test Site are reached by travelling a multilane highway northwest from Las Vegas for a distance of about 100 km. The Test Site served as the United States primary nuclear weapons testing facility from 1951 to 1992 with the government testing more than 900 above- and belowground nuclear devices there and where some form of nuclear testing research continues today. The town of Mercury is located on the Test Site a short distance inside the main entrance and serves the needs of the workers, supplying housing, a cafeteria, offices, workplaces, warehouses, a post office and recreation. The Test Site facil- ity was first operated and managed by the United States Atomic Energy Commission and currently by the Department of Energy National Security Administration. From its inception, activities at the Test Site were a focus for anti-nuclear senti- ment and within a few years of its establishment, anti-nuclear and pro-peace demonstrations began taking place along the highway route from Las Vegas and at the entrance to the facility itself.

For several decades, protesters from the United States and other countries have come to this place in the Nevada desert. They congregate and camp on undevel- oped and barren public land south of the main entrance, across the highway from the facility. This land, owned and managed by another federal agency, is rock-strewn and rough, a desert, with small and narrow flat ridges interspersed by shallow drainages. Vegetation is sparse, primarily limited to the ridges, and consists mostly of sage brush, yucca, cactus, and the odd shrub and forb. No trees are present for shade, other than the Joshua tree yucca; and there are no sustainable resources such as water within 30 km other than on the Test Site itself which was of course out-of-bounds.

Initially this gathering area was known as the Protesters' Camp and then, in the 1980s, the protesters officially named their site the Peace Camp. This location has been a meeting place and base camp for individuals and for over 200 groups with different and coeval environmental and social interests, including pacifists, anti-war groups, anti-nuclear coalitions, environmentalists, and the Western Shoshone tribe – the traditional owners in this area. The individual protesters as well as the group participants come from all walks of life, convening at the camp to present their views and feelings in opposition to local and world events. When they come together at the camp, they form a short-term, loosely organized social group for periods of short duration with the unifying focus of expressing themselves by public actions at the entrance to the Test Site and by symbolic gestures in support of peace and protesting against nuclear testing and nuclear arms in the world. The nature of the camp reflects their short-term social activities, and to some extent, their marginalized relationship to society as a whole.

Archaeology

Prior to our archaeological research, there was little information available about the camp itself. Tents and vehicles could be seen in the area during protest events, but news reports and other records focused on the events and not the nature of the encampment. During a brief visit to the camp by the archaeological team, the desert appeared undisturbed except near a 1960s' gravel pit where one could see evidence of recent occupations, such as sweat lodges, camping areas and several stone hearths. A few peace symbols made out of rocks were observed on a slope above a drainage gully and stone piles could be seen on top of a hill in the distance. The archaeological team estimated they would find 50 or so features at the site and some associated debris over an area of 40–50 ha. However, the desert environment can be deceptive and the predictions of the types and frequency of cultural materials in this setting were erroneous.

The methodology for the archaeological research was straightforward. Systematic survey was conducted with each feature or artefact numbered, measured; photographed, and the location recorded with a global positioning system. Two field sessions were conducted revealing that Peace Camp covered about 240 ha, stretching some 2,000 m east–west along the highway and about 1,000 m south from the highway. The site is not a small area with some campsites and a few pieces of art; instead, it is extensive and very complex with 771 cultural features recorded by the end of the 2002 fieldwork.

Features and Artefacts

The features at the Peace Camp are reflective of the environment and the nature of the occupation.

Most features are built with stones taken from the surrounding terrain or, in a few cases, certain types of rock were brought by someone for the creation of a specific piece of art. Stone features include rock cairns (piles), rock caches, rock circles (Fig. 15), rock foundations for statuary or sculptures, geoglyphs (symbols made of stone), rock lines along paths, rock lines enclosing an area containing desert plants (creating 'gardens'), hearths and stacks of rocks usually three to five stones tall. Sometimes in conjunction with the stone features and other times not, a flat area in the desert was scraped clean of rocks, even small ones, to create a clearing for a tent pad or for sleeping under the stars. Wood items are sparse and were imported. Logs were brought for fires and tree limbs to build structures and crates and tables for camping-related activities. Wood artefacts were scarce with most notably a wooden peace symbol and a wooden ankh. Metal artefacts were rare with most found at campsites; hearth grates with a few metal artistic objects were found at other locations on the landscape. Other features are concrete statuary foundations, barbed wire and field fencing, prayer poles, graffiti, dirt paths, dirt roads, various sculptures and symbols, sweat lodges, masks (Fig. 16), statues, willow branch structures and a porta-loo.

Fig. 15 Stone circle, one of many in the barren landscape of Peace Camp. *Photo*: Harold Drollinger

Fig. 16 A mask, one of several, all of which were found inside stone circles. *Photo*: Harold Drollinger

There are artefacts at many locations throughout the Peace Camp. The types of artefacts and their locations, such as at a memorial garden, a ceremonial fire pit and stone cairns, indicate that most have been purposefully placed at features or at special places on the landscape, as offerings or an intentional statement. Examples are crystals, a dream catcher, knives, sea shells, ceramic masks and a watch. Discarded items are rare with only a bottle or two, and small items, such as nails, a cigarette lighter and a child's toy, all probably overlooked and left by mistake. The almost total absence of rubbish is striking. Walking for hours throughout the desert, the cleanliness of the area is noticeable, especially with the knowledge that thousands of people have visited or camped there. The fact that rubbish was collected and removed from the camp in an organised manner is an indication that a set of unwritten rules or expectations existed for the protesters.

The Site

There are five focal areas at the site: an old camp, a new camp, Pagoda Hill, the highway drainage tunnels and the entrance to the Nevada Test Site. The old camp is just south of the highway drainage tunnels and west of the new camp, and the name, Peace Camp, written with aligned stones, greets anyone entering the old camp. The camp was easily accessed by two dirt roads parallel to the highway.

In the area closest to the highway, tent pads, sleeping areas, hearths, stacked rocks and rock cairns are common and scattered across the landscape. Of interest is a rock memorial garden dedicated to Ben Linder, an engineer and activist killed by the Contras in Nicaragua in 1987 (Kruckewitt 2001). People have placed small objects at the garden, probably as a tribute to him and his sacrifice.

Heading south near the Ben Linder garden is a very distinct path, its sides defined by lines of rocks. To the side at the beginning of the path is a small rock circle with lines oriented in the cardinal directions and in its centre is a posthole that once held the prayer or flagpole for the old camp. Small offerings also were left here in the stones. Alongside the path are several rock symbols, including a snake. As the path ends, it climbs onto a low ridge and one encounters a rock ring and hearth that are not habitation features. Instead, in this setting where there are no campsites, the hearth and rock ring appear to be for ceremonial use.

Further north along the ridge, outlined with stones, are a heart, peace signs, a dove and the initials TTW. At first the TTW seemed enigmatic and out of place, but as research progressed, the initials made sense as a tribute to the prominent environmentalist writer and activist, Terry Tempest Williams. She has been a participant in demonstrations at the Test Site and her concerns for the environment have been expressed in a strong voice heard by many. As a citizen of Utah, she is also a member of a group of people known as the 'downwinders', people who lived downwind of the Test Site and were in the path of fallout from some of the atmospheric tests (Williams 1990).

During the mid to late 1980s, this camp took on a different aspect when at least two residential trailers were hauled into the area. The protesters had decided to make their presence here a permanent feature of the landscape and at least one person lived permanently in one of the trailers with others staying for shorter intervals. In 1989, the government evicted and arrested some of the residents and removed the trailers from the old camp because of the illegality of residing on this type of government land on a permanent basis. Protestors promptly moved the camp eastwards to the public rally area adjacent to the main entrance of the Test Site (Cohen-Joppa 1990). Today, the only indication of this occupation is an area of disturbed soil southwest of the old camp.

After the forced abandonment of the old camp, activity shifted to the east and the protesters began using the new camp as their primary activity area. One advantage of this new location was that it had direct access by way of a highway slip-road and underpass to the entrance of the Test Site. This made the new camp more visible on the landscape to Test Site workers and those driving by on the highway.

A dirt road leads from the slip-road through a fence, south through most of the camp. East of this dirt road, individual campsites are common and there are a number of rock rings, hearths and tent pads. Standing out on the landscape is a porta-loo, donated by the Department of Energy for the new camp as a goodwill gesture. It was perceived, however, as 'something of the enemy' and vandalized by filling it with large rocks and rubbish making it unusable. In this condition, the porta-loo located towards the middle of the camp took on a new role and became a symbol of action against the federal government.

At the south end of the road is a ceremonial area dominated by two sweat lodges and a large, stone fire pit. Also present are materials to repair and rebuild the sweat lodges, rock symbols of a flower and cross, and structures made of willow branches. Between the ceremonial area and the fenced entrance, directly west of the road, is another and relatively smaller ceremonial area with a large geoglyph of a circle with rock lines pointing in the cardinal directions like a compass. Nearby is a prayer pole that is used as a central feature in the Western Shoshone sunrise services. The area probably was used for this and other ceremonial activities. Adjacent to these is a sweat lodge centre hole, the sweat lodge itself having been removed.

From the old and new camps, paths lead to the southwest corner of the site, and a hill, called Pagoda Hill by the protesters. The main path to the top of Pagoda Hill is on its north side, and stacked rocks are frequent along the route, guiding the traveller to the top. Protesters have journeyed to the top of this hill for years. Dominating its crest are three rock cairns; two are over two metres tall. These cairns or stone piles were created by protesters carrying a rock to the top on each of their visits. Offerings placed on and inside the cairns include yarn, sea shells, white quartz rocks, sandstone rocks, a Jamaican dollar, clay cherub, green stone, sage bundle, bell, white-handled pen knife, pebbles, cactus branches, an amulet, silk scarf, necklaces, tarot cards, model of a dolphin, Zia Pueblo sign, tortoise shell and notes in containers. Also, atop one of the cairns is a large quartz crystal. Between the cairns is a pole with arrow designs on its east and west sides and engraved with the words 'Healing Global Wounds', and 'May Dignity and Peace Prevail'. There is a scatter of small pieces of white quartz in this area and nearby on the crest is an arrangement of white rocks arranged as a compass. Also on top of the hill is a basalt peace symbol. On the west side of the hilltop is a red clay sculpture of a female, lying on her back on the ground (Fig. 17).

She appears pregnant and her body is covered with radioactive symbols. Hanging around her neck is an amulet with the words, 'DOE Nuke Waste Dump'. Pagoda Hill is the highest location within the Peace Camp, and from the top of the hill is a commanding overview of the surrounding terrain including the south end of the Nevada Test Site and the town of Mercury. All indications are that Pagoda Hill is a ceremonial location with the journey to the top an act of pilgrimage.

In contrast to the top of Pagoda Hill and the openness of that setting, are the tunnels that were built under the highway for drainage. Concrete-lined, they provide respite from the sun, an access route from the old Peace Camp to the Test Site boundary, and more significantly, a place for protesters to express their feelings artistically (Fig. 18). The interiors of the tunnels are covered with graffiti – literary quotes, images, abstract designs, protest sayings or chants – that illustrate their viewpoints or signify who they are or what organisation they represent. These graffiti represent the people at the site; they identify them, just like the rock-aligned symbols left on the surface of the surrounding desert.

For most protesters, their destination point is the Test Site boundary and worker's entrance to the facility. During the protests, the participants walk north from the Peace Camp, passing under the highway, and then onwards to the boundary of the Nevada Test Site. Placards portray their concerns and occasionally they obstruct

Fig. 17   The female figure on Pagoda Hill (detail). *Photo*: author

the flow of traffic. At times, they walk onto the restricted facility, resulting in their arrest. The Test Site boundary line used to be delineated by a cattle guard that recently was replaced by pavement with a wide white line across it.

North of the entrance and inside the Test Site is a trailer for security personnel and fenced holding-yards for the protesters detained by the sheriff's department. A public area immediately south of the gate and boundary line contains hearths, rock cairns, stacked rocks, and ephemeral rings in the ground, the result of dancing at sunrise ceremonies. Still tied to the wire on the fence line that demarcates the Test Site boundary are remnants of cloth placed there by the protesters during the demonstrations.

These five focal areas of the camp are only a part of the site. At first glance, the rest of the Peace Camp looks as if it has not been used by the protesters, but in reality the desert contains hundreds of features carefully placed throughout the area and which are visible only when one walks carefully across the land-scape. Stacked rocks and rock cairns are most common, and there is an abun-dance of symbolic art. This art most often is on flat land surfaces and ridges between the small drainages that cross the site. As with most of the features, this symbolic art is usually placed on the surface with a few slightly embedded into the soil; rock materials obtained locally are the predominant artistic medium. Some of the symbols are recognisable, such as peace symbols and spirals; others are enigmatic, such as stone platforms or floors. There are even large stone circles with ceramic and metal masks placed in the cardinal directions. Several are so different that they are of special interest, including a relatively large flower abstract and sculptures of children, known as the shadow children. On one small ridge, and within an oval configuration, the word 'peace' is written in English, French, Russian and Japanese, the languages of the countries with nuclear weapons in the 1980s.

Discussion

The archaeology of the Peace Camp is the archaeology of mostly non-violent dissent and activism. The campsites provide documentation of the intermittent aspect of the occupation, but the symbols throughout the desert portray the purpose of the protesters. The predominance of peace symbols, flowers, doves and hearts, created on the desert landscape reflect the protesters' goals of world peace and healing 'Mother Earth'. The offerings at various locales are other visible expressions of the protesters' personal commitments. The graffiti in the tunnels are different from the symbols and artefacts in the desert (Fig. 18).

Fig. 18 Graffiti in the Tunnel of Love. *Photo*: author

The graffiti contain statements of peace and harmony, but the writings and scenes also portray the anger and frustration of some of the participants, aptly placed inside the tunnels and not out in the open on the earth.

The camp itself is material evidence of social reaction to nuclear testing that has grown to encompass broader environmental and cultural issues, such as Western Shoshone rights and views. The Western Shoshone under the agreements in the nineteenth-century Ruby Treaty between the Western Shoshone and the United States government continue to lay claim to the Nevada Test Site land and are concerned with healing the test effects on Mother Earth. Their influence and involvement is shown by the sweat lodges and prayer poles.

However, much of the symbolism at the Peace Camp reflects other traditions, modern and ancient, and the varied constituency of its occupation. The thousands of protesters from all walks of life and different countries presented a solid front against the testing of nuclear weapons. Their reason for being at the Peace Camp was a commonly held objective, the desire to bring about a nuclear-free world. For many, this goal was expressed through civil disobedience.

Protesters talk and write about crossing the boundary line as a rite of passage. Their willingness to be arrested is often a spiritual experience (Butigan 2003) and shows their commitment to their beliefs.

> I crossed the line at the Nevada Test Site and was arrested with nine other Utahns for trespassing on military lands. They are still conducting nuclear tests in the desert. Ours was an act of civil disobedience. But as I walked toward the town of Mercury, it was more than a gesture of peace. It was a gesture on behalf of the Clan of One-Breasted Women (Williams 1990: 11).

Another protester writes:

> We all lit our candles and the procession began in ones and twos down the lonely road. … the procession arrived at the entrance to the Test Site, now guarded by 20 or more police officers, … the police informed the people that those who entered would be cited for trespassing and held in a fenced area … Over three hundred of us decided to cross the line…. Upon release from the holding area and being cited, we line-crossers were welcomed back. …The welcome I received was from an elderly Japanese man who was a nuclear survivor from Hiroshima (Peach and O'Brien 2000).

Test Site workers drive daily by the camp but do not stop. There is curiosity but a reluctance to enter the space of those opposed to their activities, while the protesters seek to enter the Test Site to demonstrate their commitment to their cause, to draw attention to their goals, and in some cases to disrupt activities there.

A retired engineer summarised the Test Site workers' viewpoint well when he said:

> You have these people that go out there and sit outside and protest. I always said, and will say it until the day I die, the very thing that they were protesting against is the very thing that allowed them to protest. It gave them the freedom in this country to do anything that they want to do including protest (Beck and Green 2004: 15).

While the protesters seek an end to the activities at the Nevada Test Site, at times this outcome often seems unattainable to them. Yet, writing about the Nevada Test

Site, Terry Tempest Williams is optimistic when she talks about 'A Rock of Resistance, Stones of Compassion'.

> When ... a poet from Kazakhstan ... came to visit the Nevada Test Site in ... 1995, he initi-
> ated an old ... custom: Each person takes a stone and places it in a pile. 'It starts small,'
> [he said] 'But one day this mountain of stone will close this test site down.'... In 1995 ...
> on the fiftieth anniversary of the bombing of Hiroshima, [Terry Tempest Williams] visited
> the Test Site to pay respects with many citizens from around the world ... the rock pile
> started in 1991 had grown to eight feet in height. [She remarked that] stone by stone... this
> is a gesture of hope [for peace and the end of war] (Terry Tempest Williams 2002).

Looking back at the end of the Cold War and the role of anti-nuclear protesters, some involved have asserted that the Nevada Desert Experience and its vigils at the Nevada Test Site were critical in expanding the anti-nuclear testing movement, creating a social climate that allowed society to accept the nuclear testing morato-rium in 1992 (Butigan 2003).

Conclusion

The Nevada Test Site is significant in the history of the Cold War as a testing ground for nuclear weapons, and the world's nuclear testing locations were the only places nuclear weapons were used during the Cold War. The Peace Camp was created in response to the existence of the Test Site. In opposing the work at the Nevada Test Site, the camp is inter-related and directly connected to the facility. The material remains at the Peace Camp tell the story of those who objected to government policy and the world political situation. Together, the Nevada Test Site and the Peace Camp represent a duality of Cold War views.

In recent times and primarily because of the 1992 nuclear testing moratorium, the frequency of protests and the number of protesters have declined. This some-what subdued turnout may be viewed as a reflection of the social and political milieu of the times. For instance, according to the *Bulletin of the Atomic Scientists*, the Doomsday Clock is currently set to seven minutes before midnight, a number that is not of great concern and near average since the inception of the clock in 1947. The furthest time from midnight occurred during the 1990s when there was a significant reduction in the stockpiles of nuclear weapons for both the United States and the former Soviet Union with the minutes ranging between 17 and nine. During the previous decade in the 1980s, however, the minutes were much closer to midnight varying from three to six. The 1980s was a time when there was an increase in nuclear weapons due to an accelerated arms race between the two super-powers. Perhaps in response, it was also during the 1980s when the protests at the Peace Camp were the most frequent and intense. From 1986 to 1994 over 500 demonstrations took place involving more than 37,000 participants, 15,740 of whom were arrested. In 1988 it was estimated that 8,800 participants were involved in a single protest event, with 2,067 arrested. Although the number of participants has dwindled since the 1992 moratorium, some continue to come and regularly protest at the Test Site with larger groups participating in annual demonstrations, such as those at Easter time and on Mother's Day weekend.

The archaeology of the Peace Camp is an opportunity to understand the material remains of a twentieth-century minority political movement. The anti-nuclear activists want to be rid of all nuclear weapons to gain world peace and harmony, end pollution of the earth and honour all living things including Mother Earth, while the Test Site, as representative of the government, seeks to gain stability and peace, albeit an uneasy one, through the strength of the nuclear weapon. Each side has its monuments and symbols. The ones at the Peace Camp are made mostly of stone, are relatively small and simple, and individualistic. On the Test Site are various industrial complexes scattered across the facility, built of concrete and metal. Remnants of past nuclear tests dot the landscape, with a few towers, remaining as symbols of testing.

The Peace Camp was and continues to be active concurrently with the government power structure that is the focus of the dissent. Instead of engaging in acts of destruction to express their desires, the people at the Peace Camp have put their efforts into creating symbols in the desert as testimony to their intent and hopes, establishing their own, separate permanent cultural legacy.

Chapter 6
Twyford Down

Twyford Down (Hampshire, UK) is one of the better preserved areas of chalk downland landscape in southern England, with surviving earthworks dating from the later prehistoric to the modern period. It is a perfect place to learn about landscape, and the processes of change and creation that continue to affect it. We can include in that the construction of the M3 motorway extension through the Down in the late 1980s (Fig. 19).

This project was hugely controversial, involving the removal of two scheduled monuments and a Site of Special Scientific Interest. Many consider this latest change to have been devastating and destructive, and some protested actively against it. This chapter suggests a contrary view, arguing that these modern additions to the landscape can become part of its value, reinforcing the point about change and creation, the very processes archaeologists routinely accommodate in studies of earlier periods. Why should we treat the contemporary past any differently? As individuals we may object to new roads and the expansion of car culture (as many archaeologists did in the case of the Newbury bypass in the 1990s, e.g. Graves Brown 1996), or to an airport extension, or industrial or housing development on greenfield sites, but that is not the same as giving recognition to those changes once they have occurred, to the process of change itself and to its physical manifestation. Following a review of some of the main issues concerning Twyford Down and a description of the area, this chapter takes one specific component of this contemporary landscape – a chalk monolith erected by pro-environmental protestors – to discuss the value of, and question the lack of wider support for this more contemporary post-modern and multi-vocal view of landscape.

Issues

In a publication that discusses the significance of later twentieth-century landscape (Bradley et al. 2004), interest in the contemporary past is explored and assessed through the related fields of historical archaeology, heritage, material culture studies and social anthropology. It recognises landscapes of the later twentieth

J. Schofield, *Aftermath: Readings in the Archaeology of Recent Conflict*,
DOI: 10.1007/978-0-387-88521-6_7, © Springer Science+Business Media, LLC 2009

Fig. 19 General view of Twyford Down, showing the old Winchester bypass (*foreground*) and the
M3 extension under construction. *Photo*: English Heritage

century as being significant, powerful and often contested. It promotes the idea that
the process of change is in itself interesting, whether or not the landscape that
results is deemed by everyone to be significant.

Twyford Down sits comfortably within these agenda. It is a landscape from
which people can examine the many values that different constituent and cultural
groups attribute to it, and about the process of change; they can learn about the
impact of major works on the landscape – their physical, emotional and psycho-
geographical influences on people and the places where they live. Twyford Down
can serve all of these objectives, but perhaps it would do so more effectively were
the more obvious signs of twentieth-century influence, and especially material
culture related to direct action and environmental protest, to be retained. Places
such as these can be defined as being discordant rather than harmonious – sites that
promote a dialogue, a debate that might not otherwise occur. Dolff-Bonekaemper
(nd) describes such places as having 'discord value' (from the German *Streitwert*).
One of the components of Twyford Down is a chalk monolith raised by roads'
protesters (see below) (Fig. 20).

Should this be kept, moved or destroyed? Each of these options has been
suggested. For example, it could be moved from its current location above the M3,
perhaps even to the National Motor Museum at Beaulieu, a place whose sole con-
cession to the negative social impact of the car is described by Graves Brown
(1996: 26) as a small, dated display on the safety of motoring (there is no mention

Fig. 20 The standing stone above the M3 cutting. *Photo*: author

of pollution or the destructive effects of road building, despite its central importance over the last 20 years). But in terms of cultural heritage, would removing the
monolith from its context reduce the impact of the important point it makes about
continuity: that this road is the latest of many, at least three of which – a Roman
road, medieval 'dongas' (trackways) and the infilled bypass – are now truly
'archaeological'? Of course, none of these options is politically neutral, which is
where the difficulty comes in.

Before discussing the monolith in more detail, it is necessary to describe the
contexts that together give it significance and meaning: these are first, archaeological
and second, political.

Archaeological Context

Experiencing Twyford Down

Twyford Down lies immediately to the south-east of Winchester, a popular historic city in the county of Hampshire. Prior to the construction of the M3 extension in the late 1980s, Winchester's bypass, the A33, ran through a narrow corridor between the Itchen Navigation canal and medieval and post-medieval water-meadows to the west, and the truncated slope of St. Catherine's Hill to the east (see Fig. 19). At the south end of its cutting were the 'Hockley Lights', a notorious set of traffic lights that routinely caused traffic queues and sometimes serious tailbacks in each direction (see Fig. 21a).

In addition, the narrow and bending two-lane carriageways that comprised the Winchester bypass caused frequent accidents, partly due to the lack of verge, and partly the sheer weight of fast-moving traffic, with frustrated drivers free at last of the long jam. In other words, this was a significant emotional and psychological landmark for anyone travelling by road to or from central southern England. Most people who travelled this route in the later twentieth century will have experienced the frustration of Winchester, and seen Twyford Down through their car windows.

Fig. 21a Protestors at the Hockley lights. *Photograph* by kind permission of the *Hampshire Chronicle*

Fig. 21b   Protestors in the M3 cutting. *Photograph* by kind permission in the *Hampshire Chronicle*

Bronze Age to the 1940s

In cultural heritage terms, the Iron Age earthworks on St. Catherine's Hill dominate the scene (Hawkes et al. 1930), being a large univallate hillfort on a steep-sided chalk hill overlooking the Itchen Navigation canal and water-meadows to the west, the city to the north-west and Twyford Down to its south-east. The defences completely enclose the rounded hilltop forming an oval area of some nine hectares. Excavations by Christopher Hawkes in the 1920s (ibid.) demonstrated the presence of an unfortified Iron Age settlement (550–450 BC) prior to the construction of defences in 250–200 BC. Evidence for earlier and later use of the hillfort includes, at the centre of the hill within a wood, what may be a large Bronze Age burial mound, while nearby are remains of a medieval chapel, built prior to the middle of the twelfth century (ibid.). In this area are associated medieval earthworks including boundary ditches, rubbish pits, chalk extraction pits and a possible cemetery. A ditched enclosure north of the hillfort and a woodland enclosure to the east are also probably medieval in date. Evidence for post-medieval use of the hillfort includes a small turf-cut maze (known as a mizmaze) thought to have been originally cut between 1647 and 1710, before being recut to a different pattern between 1830 and 1840. This is a focal point for visitors to the hill.

Twyford Down lies to the south and east of St. Catherine's Hill and, until the M3 was built, it was possible to walk freely from one to the other, though having first crossed or passed under the Winchester bypass if coming from the town. The main features of Twyford Down were: later prehistoric and Romano-British settlement, field systems and burial monuments, investigated first in 1933–34 (Stuart and Birkbeck 1936) and later excavated in advance of the M3 construction (Walker and Farwell 2000); and the medieval trackways and Roman road, parts of which survive north-east of the M3 cutting. These trackways form a well-defined series of linear ditches probably representing medieval drove or cart routes into Winchester. The trackways took on particular significance during the M3 construction, their name – the dongas – being adopted by environmental protesters. For this and other reasons, they have particular significance in understanding the continuity of use and the strong sense of traditional ownership in this area. Both the dongas and the Roman road are situated together on a chalk spur, merging from the south-east to virtually a single lane as they approach Winchester.

The industrial and post-medieval periods can also be recognised through their physical remains. In the dry valley south of St. Catherine's Hill are three burial mounds, being plague pits marking common graves of 1666 (Walker and Farwell 2000: 4). Immediately at the base of St. Catherine's Hill to its west is the Itchen Navigation, canalised in the late nineteenth century, and one of the earliest canals in England. The water-meadows further west are some of the best preserved in the county. Visible here is an entire set of water-meadows, being an area of pasture within the Itchen valley deliberately flooded (floated) to encourage the early growth of grass for grazing. Finally is the Hockley railway Viaduct, to the south-west of the hillfort and crossing both the river, the Itchen navigation and the water-meadows. Originally called Shawford Viaduct, this was opened in 1891 and used until 1961 for passengers, and 1966 for freight. It was heavily used to transport troops and war *matériel* in both world wars, and particularly in the D-Day preparations.

Late Twentieth Century

This diversity of well-preserved monuments is invaluable for teaching landscape archaeology. No one questions that. My difficulty is with those who consider that these earlier field remains *alone* constitute the cultural landscape of Twyford Down. To my mind there is much more to it. Archaeologists are now increasingly familiar with the study of the contemporary past – that which we ourselves have created and experienced (Bradley et al. 2004; Buchli and Lucas 2001; Graves Brown 2000). Equally, archaeologists are used to dealing with change, to recognising the re-use of earlier monuments or their adaptation or extension, and the creation of new ones (Bradley et al. 2004); and we now recognise the significance of new landscapes of transport, retail, science and information technology for example (Penrose 2007). At Twyford Down, a clearly late-twentieth-century contribution to landscape is evident in the construction of the M3 extension, for which some brief political context is necessary.

Political Context and Continuity of Use

There was fierce debate about the M3 extension, about where it should go, how it could be accommodated and afforded and even whether it was needed at all (Bryant 1995). Should there be a cutting or a tunnel? Should the existing bypass be extended westwards into the nationally important water-meadows? At the time, these were considered important only for ecological reasons, but now, their archaeological value is also recognised. Eventually these matters were drawn to a close. In the *Daily Telegraph*, 24 April 1989 a report described how 'Inquiry backs M-way through Bronze Age village'. It stated that:

> In what is likely to be one of the most controversial planning decisions of the decade, a [Public] inquiry inspector has recommended that permission be given to build an extension of the M3 through one of the most heavily protected landscapes in the country.

> If confirmed by Mr Channon, Transport Secretary, the route outside Winchester will mean a cutting 100ft deep and 400ft wide through two scheduled ancient monuments, an area of outstanding natural beauty and one of the last habitats of the Chalk Hill Blue butterfly.

> Conservationists say that a decision to override so many nationally important protected areas, including a Bronze Age village and field systems and medieval and Iron Age track-ways, would undermine the Government's credibility on environmental issues.
> The inspector who conducted the public inquiry on the Bar End to Compton section of the M3 concludes 'with great reluctance' that a proposal backed by conservationists for an £80 million tunnel under St. Catherine's Hill and Twyford Down, the last unspoilt hills around Winchester, has difficulties and is not justified in terms of delay or cost.

The Public Inquiry noted that the:

> dual three lane motorway some 4.2 km long would continue the M3 from Bar End to curve generally from [the] south west … to cross Morestead Road before proceeding through Twyford Down in a deep cutting east of St. Catherine's Hill (Public Local Inquiry Document File RSE M3/5/61/2/1, pp. 139–140).

This chapter makes no judgement on these matters, though they do clearly contribute to our understanding the significance of these modern components of the cultural land-scape, as monuments to change, to road building, transport, conservation and public opinion. Standing on St. Catherine's Hill today, one can see to the west the line of the former 1920s' bypass, one of the oldest stretches of dual carriageway in the country (www.cbrd.co.uk). This was backfilled once the new road was open in the early 1990s, and where the base of the hill had been cut away to make room for the road, the profile was reinstated. It is noteworthy that a section of this reinstated road has now been re-excavated once more to accommodate a park-and-ride car park, actions that once again caused protestors to return to the site. But this area at the north end of the cutting apart, trees have now been planted and in time all visible traces of the former road will be lost. For now it remains obvious in changes of vegetation that can be traced along its length.

If this infilled road is considered by some the environmental 'gain', the 'loss' can be seen to the east of the hillfort. As we have seen, prior to the M3 construction, one could walk uninterrupted from the hillfort across onto Twyford Down. That is now possible only by using a footbridge decorated with pro-environmental slogans (Fig. 22). Those currently legible include:

Fig. 22 Environmental slogans above the M3 cutting. *Photo*: author

'Smash the D.O.T.'s [Department of Transport's] new roads'

and

'No child asthma or earth rape Mr. Malone'.

Another damaged piece of graffiti states:

'The destruction of Twyford Down [by] Mr. ...'.

One could argue that, as a cultural expression, this approach to Twyford Down – across busy traffic lanes – is more stimulating, more challenging and more informative than a stroll across the Down could ever be.

Also on the bridge, small stickers containing protest slogans have at some point been stuck to it and a few remain – subtle traces that nevertheless contribute to the interpretation of the place in recent memory.

And then there is the M3 motorway itself. Excavations in advance of the M3 construction involved the removal of prehistoric, Romano-British and medieval archaeological remains (Walker and Farwell 2000). But it also created a new layer to this cultural landscape: a new monument for visitors to interpret and understand.

Taken in isolation, this road perhaps is not hugely significant. There are many new roads, all of which are relevant in documenting late twentieth-century politics and political (re)action: the Newbury bypass for example, the M11 link or the A30 in Devon (Butler 1996). But here there is continuity, and that is partly what sets this site apart. In a closely confined area is the Roman road, medieval trackways that may have earlier origin, the early bypass now backfilled, the modern M3 extension and the monolith (described below). Additional to that is a strong connection between the landscape and the environmental protestors who camped here to oppose the M3 project. These protesters considered themselves a tribe, taking a large part of their identity, including notably their name (the Dongas, as we have seen the local toponym for the medieval trackways running up the hillside) from the landscape they were trying to protect. They are still present in the landscape in the faded environmental slogans adorning the bridge across which everyone now wishing to walk between the hillfort and the Down must cross. It was this tribe of some 40 protestors who settled on the Down that provide the most obvious connection between political action and landscape and it was they, and specifically their activism at Twyford Down, which sparked the anti-roads protest movement in Britain and national environmental campaigning. As one member of the Dongas told a reporter: 'Call us indigenous Albion, if you like. We have chosen this. We are passionate about Life' (Anon. 1993). That was their claim, and their philosophy, and who are we to question that?

The Chalk Monolith: To Those that 'Ravaged the Land'

The monolith (Fig. 20) is located immediately east of the M3 cutting, at its northern end and in an area of public access. It is situated within a small clump of hawthorn trees in which fabric offerings are sometimes placed. In June 2004, a spiral arrangement of small pebbles was visible in front of the stone. The monolith is an upright 1.5-m-high piece of chalk, believed to have been taken from the cutting, inscribed as follows:

> This land was ravaged by:
> G Malone
> J MacGregor
> R Key
> J Major
> D Keep
> C Parkinson
> C Patten
> M Thatcher
> C Chope

All those named had a significant part to play in the M3 extension, and road-building and environmental policy in the late 1980s. Two former Prime Ministers are included in this list, as well as numerous senior politicians and local councillors.

At one level, the monolith represents a physical reminder of the reasons behind the current extent of surviving pre-road archaeology. There is a cogent argument also for its significance in terms of association and continuity, particularly as it overlooks the dongas, the M3 cutting and the footbridge with its pro-environmental slogans. While many would understand and accept the cultural relevance – even significance – of this monument, a suggestion to give it official recognition as part of the protected (scheduled) area at Twyford Down proved problematic and controversial.

In 1999–2000, English Heritage sought to revise the area of Twyford Down and St. Catherine's Hill that was already afforded statutory protection as a scheduled monument under the terms of the 1979 *Ancient Monuments and Archaeological Areas Act* (see Chap. 1 for a discussion of what this means). As is usual, this process of revision involved the site being visited and assessed by an English Heritage archaeologist who then produced a report for submission to the Department of Culture, Media and Sport (DCMS), the government department responsible for heritage matters in England. English Heritage in other words provided the advice or the recommendation; the DCMS would make the final decision. During this process, the English Heritage review considered the monolith and decided to include it within the proposed new scheduled area. At the time, this was perhaps the most recent monument ever to be considered for scheduling, and it was also one of the most inherently politicised. Given that the monolith met the legal definition of monument (there is no time limit defined in the *Ancient Monuments and Archaeological Areas Act*, and it is a 'building, structure or work'), and that it fell within the area of the recorded extent of medieval trackways, how then should DCMS treat the monolith? Either to include it or specifically exclude it from the scheduling would require justification. It was not possible merely to ignore it. Consequently, whatever the DCMS decided could be interpreted as a political act, given that by this stage it would be a Labour government giving (or denying) national status to a monument decrying the actions of a Conservative one. As English Heritage stated at the time:

> Excluding the monolith [from the scheduling] is not a neutral decision. It could carry a message that some aspects of the historic environment are not granted significance by English Heritage because they are too ephemeral or modern or because they exclude more widely from official perspectives of the recent past. It would encourage accusations of sanitising and distorting the past, and of creating an approved heritage. On the other hand, the monolith's inclusion might offend other sensibilities, not least of course those named on it. … In our view, however, inclusion can be defended as the most neutral decision on the grounds that the monolith is a physical reminder of the reasons behind the current extent of pre-road archaeology [and] on grounds of association and continuity (Fairclough and Schofield 2000).

English Heritage also noted that:

> this approach would be consistent with current initiatives on cultural diversity and social inclusion, and the attempt to re-negotiate the concept of heritage with social, community and interest groups other than those normally perceived as being part of the conservation and heritage debates (ibid.).

Against this background, would the monolith be retained through statutory protection, recognising its significance within the wider changing landscape?

Views

Although responses were inevitably biased by experience and cultural and intellectual background, a questionnaire was completed by first year archaeology undergraduates at Southampton University as part of their Landscapes and Monuments option. In March 2004, 29 students visited St. Catherine's Hill and

Twyford Down, spending time examining the hillfort, mizmaze, chapel, former bypass, M3 and the monolith. Afterwards, they answered the following questions (amongst others):

1. Should the chalk monolith be preserved or not? Give reasons.
2. Why and for whom does the Twyford Down cultural landscape have significance?

The answers were generally eloquent and strongly expressed. Taking here the first question, responses included references to the monolith being a 'recent addition to the continuous story of the landscape'; a 'reminder that other voices were heard'; and that it 'continues the story of change and development in the landscape'.
Also:

> Although I do not believe the road should be there I still think the monolith should be preserved as it is still a piece of history and is now part of the landscape. It represents how people felt because of the road's construction, and will give people in the future a good indication of the outrage it caused.

> The monolith is as much a part of the history and issues of the site as anything else and the people who created it have a right to have their opinions made visible.

> Despite the fact that I do not agree with the monolith or its message, I believe that people have a right to oppose government decisions and moreover people have a right to see that the road was opposed. To preserve [the monolith] would not be a hostile action towards the government [of the time] but would ensure that people's views are listened to.

> Archaeology should be a neutral subject, thus politics shouldn't be included in scheduling decisions. The monolith is now part of the landscape; it says something about the current state of the landscape and what can be preserved for future generations, so they can decide how they feel about it.

So what happened? This was one of those rare occasions when English Heritage's advice was not taken by DCMS. The Department decided that they would extend the scheduling of St. Catherine's Hill and Twyford Down as recommended, but specifically *exclude* the monolith because it was considered too soon to put the structure – and the significance of the protest – into a proper historical context. But for now the monolith remains as a legible and obvious feature, representing the latest stage in a history dominated by the need for transport and communication either to and from Winchester or – uniquely for the twentieth century – that avoids it. It is inevitably and unquestionably part of this landscape, albeit without the protection that could ensure it remains, allowing future generations to consider it and to decide its relevance.

Conclusion

This chapter has considered two related issues within the general and related themes of cultural heritage and modern material culture: First, that cultural heritage does not constitute a static, unchanging record of human impact on and interaction with the land. Landscape is constantly being renewed, changed and reinvented, to the extent that much of what we see today is characteristically post-1950s' landscape, reflecting modern and contemporary socio-economic trends such as housing, militarism, transport and leisure. Indeed it was some of these trends that provided the justification not only for Winchester's M3 extension, but also for many of the recent road schemes that have caused such controversy, and created such impact at landscape scale. These 'monuments' therefore are inevitably the key sites at which these trends can be assessed, reinterpreted and ultimately questioned in the future. One could see this as entirely negative, as change that has destroyed the idyll that existed before. But there is a more positive interpretation, of a changed landscape containing the evidence for a diversity of human activity and cultural processes from prehistory to the present (see also Penrose 2007).

The second point concerns the significance of this place in particular. Some places have cultural and social value for what they represent in terms of achievement, or key political actions and events; for representing defining moments in history on the one hand, and broader socio-political trends on the other. Greenham Common is one such place, and Twyford Down is another, not for the reasons that led to its protection as a scheduled monument, and the revision of that designation, but for reasons that have been deliberately *ignored* in the scheduling. Perhaps the decision not to include a recent monument within the scheduling may be reversed in time, depending on whether it was an unwillingness to enter a political argument, or whether it really was unfamiliarity with modern heritage and the ways in which we assess it that caused it to be ignored. Or perhaps lack of designation can be seen as appropriate? How can one make the discordant official without affecting its character? It was never suggested to schedule the road, mainly because it has a future in everyday use. One could argue that the monolith also remains in use, both physically and symbolically.

Twyford Down is a special place (and in a recent opinion poll it was voted the most spiritual place in the region), but it is the more recent changes to this landscape that give it the edge as cultural heritage. The old infilled bypass, the new cutting, the monolith and the offerings left there, deciphering the faded slogans on the bridge, and the remnants of protest stickers, all combined with the earlier remains, the sounds and the smell of the traffic, the birds, and kids enjoying the mizmaze on a summer's evening – all contribute to a multi-sensory, multi-vocal, multi-period and characteristically post-modern experience of place. That's why Twyford Down should matter, to English Heritage, English Nature, the local community, its owners, Indigenous Albion … everyone.

Chapter 7
Greenham Common Airbase

With Mike Anderton

> Imagine a bomb up the bum of suburbia. But the bomb is made of organic flour, wrapped in ivy, painted in funky colours and thrown by pixies; half punk, half pagan. The spirit of the direct action protest movement is like this, half 'spiky', half 'fluffy' – half politically hard, half warmly, humanly, soft. The movement boils with life lived to the brink, to the full, its emotion intense, raw and extreme (Griffiths, in Evans 1998).

Like Chaps. 5 and 6, this chapter also concerns opposition and protest, and how the materiality of opposition provides a necessary contribution to achieving a full and balanced interpretation of past events and social actions. Unlike Butler's (1996) survey of cultures of resistance on the M11 Link Road, the monuments of which were destroyed in the road's construction, this chapter further examines what survives materially both of the actions of those who protested against nuclear armament during the period of the 'Second Cold War' (Hobsbawm 1995: 244), and of the targets of those actions, and how that material culture should be presented to future visitors to the sites. There is also the point made by Uzzell (1998) that Cold War sites are different to those of other wars in that they are often not the scenes of conflict and death; that their importance and value lie in what they represent and what might have been.

In all these senses it is an unconventional archaeology that we present here, atypical in its associations, uncertain (or at least debated and contested) in its meaning, mysterious and disquieting in its Cold War context, outlandish and unorthodox in what it can hope to achieve in terms of public perception and interpretation. Protest is the stuff of everyday life, yet recent examples are rarely and barely recognised in heritage interpretation; particularly when opposition was directly aimed at the establishment view or government policy. Here we make the point that protest in the form of direct action – violent or not – inevitably sets up a contradiction, a challenging dilemma, for heritage managers to confront, not avoid; conflicting archaeologies, if not archaeologies in conflict.

The example we use is the area either side of Green Gate at Greenham Common Airbase in West Berkshire, UK. It was one of seven gates – all named by protesters after colours of the rainbow – that surrounded the base and which each had an active and – for the Ministry of Defence, USAF and NATO – disruptive Peace Camp with a distinct mood and atmosphere. Yellow Gate or Main Gate was the largest

J. Schofield, *Aftermath: Readings in the Archaeology of Recent Conflict*,
DOI: 10.1007/978-0-387-88521-6_8, © Springer Science + Business Media, LLC 2009

camp, had all the traffic noise, and had a 'special urban desolation that made it grimmer than the rest of the camps' (Blackwood 1984: 29). The women who lived at Orange Gate were squeezed up against the perimeter fence and were therefore nearer to the soldiers and their 'terrible sexual taunting' (ibid.: 23). Green Gate – 'the camp of intellectuals' (ibid.: 7) – lay immediately outside the GAMA site (Ground Launched Cruise Missiles Alert and Maintenance Area) which came to prominence at the time of the 'Second Cold War' of the 1970s and early 1980s following the announcement that Tomahawk Ground Launched Cruise Missiles (GLCMs) would be deployed there.

The 'Greenham Women', whose members strongly objected to this deployment and to nuclear weapons and technology generally, chose 'the power of non-violence to counteract the power of evil, generated from inside the base by genocidal nuclear weapons' (Hipperson 1998: 356). Some even described the base itself as a 'nuclear concentration camp, where preparations for mass murder are carried out daily' (op. cit.). However, despite the strong emotional content of this language of protest, actions by the women were non-violent, and embraced an outlook or 'mood' close to that expressed in the opening quotation. Statements of opposition were maintained, and ideological and spatial positions were negotiated, as was the case with earlier political and environmental protests against nuclear armament that began with the Aldermaston marches in the 1950s, and spun off into a range of environmental protests at Greenham, Solsbury Hill (1994) and the Newbury bypass (1996). Of course, these are all related and, in fact, Kate Evans' *Copse: The Cartoon Book of Tree Protesting* (1998) begins at Greenham Common's Violet Gate with childhood memories of bracken sunhats and picnics by the fence. But Greenham was different, in being consciously and deliberately women only (unlike contemporary peace camps at Faslane and Molesworth, and in Nevada – Chap. 5), and its agenda was wider as a consequence. For example, the Greenham campaign had strong links with other causes such as the Prostitute's Collective, and miners' wives groups during the 1984 coal strikes. There were also the symbolic aspects of 'women's space' and protests against patriarchy, which for some women were equally, if not more, important than nuclear disarmament.

This chapter has its context in two related areas. First, there is now a general interest in the politics of non-violent protest, in a period when non-violent protest is commonplace (cf. MacArthur 1998: passim) and when interest in the materiality of the recent past is increasing (e.g. Buchli and Lucas 2001; Graves Brown 2000). Second, and more specifically, there is now a concern for the future of the GAMA site, now disposed of by the Ministry of Defence, and in particular how its archaeologies are to be presented and interpreted for future visitors, assuming that a sustainable future can be found for the monument and some physical preservation is achieved. In short, can these contrasting and conflicting archaeologies – all of which may be considered 'queer' depending on one's political and ideological stance – be presented and interpreted together in a meaningful and cogent way for the benefit, education and enjoyment of future visitors?

Cold War Archaeology and Greenham Common Airbase

> Anachronistic in normal periods, in peacetime, the bunker appears as a survival machine, as a shipwrecked submarine on a beach. It speaks to us of other elements, of terrific atmospheric pressure, of an unusual world in which science and technology have developed the possibility of final disintegration. If the bunker can be compared to a milestone, to a stele, it is not so much for its system of inscriptions as it is for its position, its configuration of materials and accessories. ... The monolith does not aim to survive down through the centuries; the thickness of its walls translates only the probable power of impact in the instant of assault. The cohesion of the material corresponds here to the immateriality of the new war environment; in fact, matter only survives with difficulty in a world of continuous upheaval. The landscape of contemporary war is that of a hurricane projecting and dispersing, dissipating and disintegrating through fusion and fission as it goes along. With the passage from molecular arms to nuclear arms, what happened in test tubes at the microscopic level of chemical and biological reactions is happening from now on in the macroscopic universe of human territory. A world of moving particles – that is the inscription of these concrete steles (Virilio 1994: 39).

Cold War archaeology is comparatively new territory for archaeologists, and specifically for those engaged in heritage management, and has been since the Berlin Wall was dismantled in 1989 (Cocroft and Thomas 2003; Dobinson 1998; Schofield and Cocroft 2007). What were until recently military installations, some highly secretive, are now being recorded or preserved as historic monuments, and are presented to a public who are increasingly aware of the significance of post-war politics, and the symbolism of Cold War structures such as the GAMA site, in shaping the modern world. They are also interested in experiencing such secret and mysterious worlds for themselves. (This interest is also reflected in historical research where the availability of documents can now ensure the publication of more informed accounts than was previously possible, e.g. Gaddis 1997.) Hence the interest in the Nevada Test Site (Johnson and Beck 1995), Orford Ness (Wainwright 1996), the USAF Airbase at Upper Heyford (Hinchliffe 1997), and the concerted attempts to have Minuteman Missile silos at the Ellsworth Air Force Base in South Dakota preserved as 'historic monuments' so that 'the Minuteman story can be told' (Wharton 1999: 48). Motivations for presenting and interpreting the GAMA site at Greenham Common have much in common with all of these examples.

The airbase at Greenham Common (Fig . 23) has a notable history (and it is interesting to draw a contrast here between attitudes to Greenham's Second World War role – consensual, uncontested and mostly approving – and it's Cold War significance, where opinions regarding the morality and the cause, let alone the means, are less clear-cut). Greenham originally became an airbase in May 1941 when the airfield opened as a satellite to RAF Aldermaston. The USAF took over, and in 1942 it was the headquarters of the 51st Troop Carrier Wing during Operation *Torch*, the invasion of North Africa, while in 1944 the base was involved in the preparations for, and the execution of, the D-Day landings. General Eisenhower, the Supreme Commander of the Allied Expeditionary Forces, visited Greenham on 5 June 1944 and gave his 'eyes of the world are upon you' speech to the men of the

101st Airborne Division; and later that year Greenham was also involved in the airlift for the Arnhem landings. After reverting to the RAF in June 1945, closure in 1946 and the land passing back to Newbury Borough Council in 1947, the Air Ministry announced its intention to re-acquire the site – the reason being the increased East–West tensions following the Soviet blockade of Berlin in 1948. The USAF took over the base again and embarked on a programme of enlargement and significant alteration; and from January 1958 until the base closed in 1964, Greenham was part of the Reflex Alert Scheme, whereby B47s armed with nuclear weapons were on stand-by for immediate take-off. The base was then de-activated in June 1964 and returned to the RAF.

The most recent period of the history of Greenham Common Airbase begins in 1968 with its reopening as a USAF stand-by base. In 1979, NATO, responding to the build-up of nuclear weapons by the USSR, decided to deploy intermediate-range nuclear weapons in Europe. In June 1980, this led to the announcement that Tomahawk Ground Launched Cruise Missiles (GLCMs) were to be deployed in Britain at both Greenham Common and Molesworth (Cambridgeshire). Construction

Fig. 23 Aerial view of GAMA. *Photo*: English Heritage

work on the new installation at Greenham began, and the Alert and Maintenance Area (GAMA) was built partly on the site of the 1950s' Strategic Air Command nuclear weapons storage 'igloos' at the western end of the runway. In July 1982, the 501st Tactical Missile Wing was activated to operate and maintain the GAMA site. The base became operational again in June 1983, with the first 16 cruise missiles arriving in November. The GAMA site was completed in 1986 and by June there were six flights of GLCMs, with a total of 96 missiles (and five spares) stationed at Greenham. On 12 December 1987, the USA and the USSR signed the Intermediate-range Nuclear Forces (INF) treaty, the provisions of which included the elimination of all cruise missiles from Europe. This meant that, between August 1989 and mid 1991, the cruise missiles were shipped out of Greenham in stages, being taken back to the USA for destruction, and in 1992, the USAF left Greenham and the base was closed.

What survive within the former airbase are solid, military remains representing all phases of its use. The GAMA site forms a parallelogram, covering an area of 495 m east to west by 450 m north to south, defined by three fences topped with razor wire, lighting and surveillance cameras. The area was (and continues to be) dominated by the six GLCM shelters (Fig. 24), arranged in two rows of three, while to the west are the 1950s' nuclear weapons storage igloos, which were refurbished as part of the GLCM deployment.

This was functional architecture in the extreme, existing only with a view to 'doing' something: waiting, watching, being watched, warning and threatening;

Fig. 24 One of the shelters for cruise missiles and their launcher vehicles, GAMA 1999. *Photo*: English Heritage (AA000532)

then acting or, rather, reacting (after Virilio 1994: 43). The shelters each had three lanes inside – designed to hold two mobile launch control centres (LCC) and four transporter erector launchers (TEL), each of which carried four GLCMs. The lanes had their own doors, front and back, that were operated by hydraulics and which, when open, covered a deep trench in front of the shelters. Each shelter had a massive concrete roof and was grass covered. Only the main – or Quick Reaction Alert – shelter, the northwestern of the group, was designed to be permanently manned and had domestic accommodation attached. The area also had a Reserve Fire Team Facility (RFTF), Missile Store, a Maintenance and Inspection building, for undertaking works to vehicles, a Control Room and Entry Control Point, with a bus stop beyond the gate at the outer fence. All of this remains, though in five of the shelters (excepting the Quick Reaction Alert shelter) the hydraulics have been removed. In all six shelters the doors are in the down position. Much of the equipment has of course been removed, with just occasional fixtures and fittings and a few scattered and generally insignificant artefacts remaining.

But that, of course, is not the full story: although less monumental, the archaeology of Greenham Common Airbase extends beyond the militarised landscape referred to here. This site, like so many other monuments, has multiple histories to be considered – histories that are not always obvious through physical remains.

Queer as in Peculiar: Life Beyond the Fence

'I was considered unusual and queer, you know, queer in the sense of the word peculiar' (in Junor 1995: 296) –- so Teresa Smith sums up the judgement of her middle-class neighbours about her alternative views on life. These views saw Teresa take part, along with her seven-year-old daughter, in the Greenham protests during the early 1980s. She was one of the many women who participated in something that, from small beginnings as a march from Cardiff to the USAF base in 1981, became a major, long-term action. Some of the participants stayed only a few hours, others were frequent visitors, for example at weekends or for special events, while others became permanent residents and remained for several years. They all left their mark on the base in some way.

At issue in this section is how archaeologists can understand ways in which these marks can be identified and worked with. An alternative form of protest, with its commensurate, 'queer' approach, has left us with the challenge of how to interpret, and engage with this queer/peculiar form of modern archaeological material. If, as Junor (1995: xi) suggests, an 'incalculable' number of women passed through the camps at Greenham, what is there to show of their presence?

We could attempt to examine the physical remains of the peace camps in order to understand what happened at the base. However, as we have already seen, the surviving physical remains at Greenham are largely those of the military base. Outside the fence there are few obvious physical reminders of the camps and their occupants. A memorial garden, stands at Yellow Gate. At Green Gate one can see the remains of painted graffiti on the road surface, on fence posts and on walls within the GAMA site; while in the woods around the perimeter slight earthworks and clearings are visible. Also, in some parts of the woods, pits and other cut features still survive, along with artefacts, as demonstrated by recent field investigations at the small camp at Turquoise Gate.

Caroline Blackwood, in writing of her associations with the camp at Blue Gate, shows us why so little remains of an obvious physical nature. She describes how, on her first visit to the camp, she arrived at night to find what appeared to be a discarded, mud-spattered plastic sheet lying nearby (Blackwood 1984: 6). On closer inspection, she realised that this sheet, and a great many other sheets of plastic, actually had women living in them. They were, in fact, a series of flimsy and easily replaceable living structures called 'benders', each consisting of several branches bent into a hooped frame and covered with plastic sheeting (hence the term). They were sturdy enough to keep the wind and rain out (most of the time); but they left no impression in the ground once they had been removed, as was obvious after every eviction by the authorities. Caroline Blackwood described her return to the camp shortly after an eviction thus:

> When I next went back there, the benders had ceased to exist. It was hard to believe those squalid little colourful dwellings had ever been there. There was now only a lot of churned up mud, and the odd piece of newspaper and the odd trampled plastic spoon.
>
> The perimeter now had an unsullied, unchallenged greyness. It looked triumphant and immovable as it reigned over the countryside with its rolling entanglements of barbed wire.
>
> But although the camp had been wiped out, a little group of Greenham women were still there. They were sitting in a circle in the mud. As usual, it was bitterly cold.
>
> They didn't speak very much. They just sat there as if they were having a make-believe picnic in mime. The food and the fire all had to be imagined. In reality they now had nothing except mud (ibid.: 79).

This last point is important as, without any prominent traces in the ground, we would be unlikely as archaeologists to know that a reasonably large group of people (let alone the fact that they were all women) actually lived near the outer perimeter of the base over a period of several years. The only substantial, remaining physical element that may be seen, and which has a consistent, easily identifiable, reference to the women and the camps, is the Base's perimeter fence itself. This physical and mental boundary acted as a continuous focus for the protesters, in a dualistic sense, throughout the time the camps were in operation (Blackwood 1984: passim; Junor 1995: passim). It was frequently cut in order to gain access to the base and its shelters (ibid.), and it still bears the 'scars' of these actions today (Fig. 25).

The fence also attained a quilt-like colourfulness on occasions too – albeit for short periods. The women took to 'darning' the fence with brightly coloured wools in a symbolic form of protest – an action through which they 'begged that they be spared from nuclear destruction so that they could still patch up the holes [that] men had made' (Blackwood 1984: 28). It was an ironic symbol that was lost on the soldiers guarding the base though, and they were quick to bring down the women's

Fig. 25 The fence separating GAMA from the peace camp at Green Gate. *Photo*: author

handiwork. In addition, the fences were often decorated with symbolic items, notably children's clothing and photographs, as well as with placards and leaflets.

Documentary and oral evidence (Blackwood 1984; Junor 1995; C. Stoertz: personal communication) also describes other symbolic actions that have left no physical trace at the site. Some women, for example, held mirrors up to the base in order to 'reflect [its] evil back into it'; and 'webs' were woven that were hung on the base's perimeter fence (Blackwood 1984: 7). These webs were not mere 'decoration' however but rather an example of the strength and unity shown by the women – a symbol of the stand against the bombs and oppressive authority at the base and around the world. They portrayed in visual form how one strand alone (that is one single protestor) may appear weak, but how many strands (the women together) were united within a more complex and less easily destructed whole (Blackwood 1984: 7 and 21; see also Junor 1994: 299). Of course there was a more straightforward motivation: to subvert the fence; to make it less male, less military, less functional ... and more ridiculous.

Though these decorations are no longer physically apparent (except in a very limited way, for example on the anniversaries of significant events), just like the benders at the camps, the knowledge of their existence leads us to recognise the role of social interaction and human agency beyond the mere physical nature of the structures and fences at the site. Students of prehistoric archaeology are now familiar with the ideas of the social nature of monuments and their surrounding landscapes, and the symbolism and esoteric knowledge that may often be associated with them (cf. Barrett et al. 1991; Tilley 1994); but can we see these same ideas at a very modern Greenham? The examples presented above suggest we can read the alternative narratives that the GAMA site presents.

There is another symbolic form to consider that is familiar to students of prehistoric archaeology, one which revolves around issues of structure and agency – that is the women's views on the nature of the land itself. If we are to observe the fence and the military base, we may, taking a pre historian's approach, be able to observe the juxtaposition between the land without and within the perimeter fence in the following terms:

public space/private space
civilian/military
female/male
life/death
hope/fear
peace/war

We may also be able to observe that the damaged fence had been subject to some form of opposition to the dominant ideology of the base through observing the attempts to break through the physical *and* mental barrier. What we do not see without the documentary and oral evidence of the women is the idea of what the land at Greenham symbolises for this group. The women came to see the land at Greenham as *their* land in a very spiritual sense (Junor 1995: 20, 25 and 55). In opposition to the dominant ideology of the base as a functioning, asocial military site, they saw a 'sacred land' being defiled by the missiles (Junor 1994: 55 and passim).

A 'sacred land'; a site occupied for over a decade; peaceful, physical protests and symbolic actions. We can see little of these issues within the currently available and visually obvious physical evidence; and this is, surely, very surprising as we are dealing with a relatively recent archaeological site. At Greenham, multiple histories abound, and we are, here, within the realm of 'incomplete narratives' (Scott 1997: 1) where, without care, we are at risk of portraying only one side of a story.

Interpretation and Future Management

Since the closure of the airbase in 1992, discussions have been taking place over its future management. Some decisions have already been taken and much work has been completed. The runways, for example, have been removed in order that the airfield can be returned to common land, as it was before the Second World War. The central part of the runways, where they form a cross, has been retained as a symbol of the site's former use and significance. One proposal concerns the conversion of the Air Traffic Control Tower for use as a visitor centre. The Tower is central to the newly recreated common, and its glass observation room will provide an ideal opportunity to view the former airbase. Three aspects of interpretation have been proposed for this facility (from ground level upwards): natural history; the history of the common; and its Cold War associations. The shelters of the GAMA site would thus be visible in the middle distance from the observation room and its Cold War interpretation facility at the top. By contrast, the area south of the runway and inside the main gate – the so-called Area E – now functions as a Business Park, many of the buildings here having been adapted for new uses.

So, with the exception of the Air Traffic Control Tower, and one or two buildings in Area E and elsewhere, much of the appearance of the Cold War airfield will in time be altered beyond recognition. We would argue that, for a site of such significance, in social, political, military, strategic and technological terms, some symbol of power and of the contradictory and conflicting stances the site represented in the later twentieth century, should be retained. The most powerful and meaningful symbol of all is the GAMA site and the archaeology of protest that existed immediately beyond its boundary, and it is this element which, we believe, holds the key to the site's future management and interpretation. A short description of how such a message of contradiction and conflict could be conveyed, and how a balanced interpretation of the GAMA site can be presented in an objective, though not dispassionate, way given pragmatic considerations of sustainability and monument management, is therefore given below.

One immediate question concerns the matter of exclusion and site security. For a site known for its exclusivity and secrecy, is it appropriate to provide open access for visitors? Should there not be some token security check to enable people to pass through the gates to the restricted area within? There is much to be said for this though, equally, removing part of the fence could serve to demonstrate the GAMA site's role in *ending* the Cold War, not just in propagating it. A compromise seems the obvious solution and would have practical benefits: in a relatively remote area, adjacent to common land, and for a site already vulnerable to vandalism, sealing it off completely for long periods and giving controlled access would prove unpopular, and may present a financial burden that the likely number of visitors could not sustain. Rather, providing open access to much of the site and some of the shelters and associated buildings, while retaining significant stretches of fence line (particularly that section from the Guard Room, around the eastern end of the GAMA site, to a point west of Green Gate) would retain the site's original appearance and atmosphere – conveying an element of menace, security and secrecy. This section of the fence provided a focus for anti-nuclear protests; it was decorated by protesters, and now has the appearance of a patchwork of cuts, repairs and counter-cuts (Fig. 25). This 'stratigraphy' tells us much about the history of opposition at Green Gate and should be retained in some form, and preferably in situ. Interestingly, cuts continue to be made, presumably by protesters as symbolic statements of opposition, actions that are likely to continue as folk tradition if nothing else.

Removing the fence line at the western end of the GAMA site would have two further advantages. First, the vegetation on the site – predominantly grass, originally mown – could be maintained by grazing, and without a full circuit, any animals grazing the common could have open access to the GAMA site from the west; temporary fencing could of course be introduced for closer stock control. Second, visitors will access the site predominantly from the east (having first visited the facility proposed for the former Control Tower). Retaining fencing on the GAMA site's eastern side would ensure those visitors experience the site as it was, passing along the fence before entering through the gate. Providing a low-key interpretation (with site plans and some photographs) in the Guard Room, serving as the essential introduction to the GAMA site, would further ensure most visitors enter this way. If transport is to be provided from the Control Tower, there is a certain attraction in using the original bus stop (complete with shelter) outside the site's main gate, and the turnstiles by the Entry Control Post for access.

So far as presenting the interior is concerned, retaining the overall character and symbolism of the site may be more important than preserving every detail of the individual structures. For example, maintaining all six shelters would be unsustainable and is unnecessary – keeping one in full condition, while reducing the others to their basic, robust form, may be more sensible. The five shelters which have already been fully de-commissioned could be stripped bare of fragile fittings in order to ensure a minimal maintenance burden. They could then be presented as robust, low mainte-nance monuments with open access (though accepting that some safety measures may need to be put in place, such as sealing off internal doors and reinforcing guard rails, measures which would themselves symbolise the act of disarmament). The Quick Reaction Alert shelter with its accommodation block could be the one example retained more completely. This was not de-commissioned and could be put back into working order. The doors could be kept closed for security and safety, but opened for accompanied tours, the opening and closing of the shelter being the event that 'sells' the full tour to those who could otherwise enjoy open access to the remaining five shelters. There could be scope for interpretation in the accommodation block, although it may be preferable to resist this, and allow such interior spaces to speak for themselves; for the present time, their significance is that they are disused and abandoned.

The level of interpretation provided for visitors is an important consideration in view of the base's multiple histories. Assuming that contextual material will be provided off-site in the interpretation facility in the Control Tower, much of the GAMA site could be left bare. Some photographs and plans in the Entry Control Post could be accompanied by limited text to describe the main components of the site, emphasising both what lay within the fence, and what was beyond it. Green Gate, leading onto the public road, could be left open, allowing visitors to walk along the outside of the fence, and to view the shelters through the patchwork of the outer fence (Fig. 25). Another possibility is to use one of the lanes in the Quick Reaction Alert shelter as an art and/or artefact gallery, if correct conditions could be ensured. Part, or all, of the Turner Prize nominated Wilson twin's video sculpture *Gamma* (Corin 1999), which investigates the themes of power, surveillance and paranoia through photographs, performance and installation art and was part recorded and filmed at the airbase, could perhaps be exhibited on occasion, as could John Kippin's photographs (Kippin 2001) and other related works. Another possibility would be to exhibit work produced by the Peace Women and their supporters. These, less official, views of the site and what it stood for could then be seen in the very place that inspired them. This is particularly relevant as alternative political views are often lacking in official heritage venues (Frank Casey's sculpture depicting a scene from the Miners' Strike in the City Museum, Stoke being a notable exception); the situation at Greenham provides a rare opportunity to redress this imbalance in some way.

Conclusion

We argue that to present the recent past at Greenham Common Airbase as it was, and not in some diluted, biased or sanitised form, is desirable but difficult. It requires an approach that welcomes opposing viewpoints (including those of local residents, and the personal views of US service personnel and their families, which have not been examined in this paper), not presenting them as some side-show or adjunct to the main attraction. In that sense, Greenham is unusual: many Cold War sites were sufficiently remote or 'secret' enough not to be affected by the actions of protesters. To present that opposition in a visual and affecting way requires the preservation of sites such as GAMA, in order that visitors can experience this significant episode in world history and international geopolitics through the unbiased and balanced presentation of views and actions. Hobsbawm has said (1995: 247) that: 'people in the twenty-first century, remote from the living memories of the 1970s and 1980s, will puzzle over the apparent insanity of this outburst of military fever [and] the rhetoric of apocalypse'; he did not say that people may not even see the opposite face, that of peace camps and protest, woven webs and mirrors – and they will certainly have difficulty recognising the full complexity of their meaning and symbolism. Heritage attractions such as GAMA, with engaging interpretative facilities, will at least give an impression of the political atmosphere in which the arms race escalated – as well as the social context of protest and opposition, and the materiality of a rather queer contradiction.

Postscript

As we completed the original paper, the Greenham Women's Peace Camp had begun to close down after 18 years of continuous work. To mark this event, the women submitted a planning application in 1999 for a 'Commemorative and Historical Site', next to the main entrance to the Airbase at Yellow Gate, to 'acknowledge and preserve [the Peace Camp's] unique part of history [and] as an inspiring contribution towards a world without nuclear weapons'. The press release, issued by the Greenham Women, stated how they:

> envisage the site as presenting the community of Newbury with an opportunity to heal the breach that has developed over the siting of Cruise Missiles. The vision of a circle of standing stones, we believe, will endow the area with a spiritual and healing influence and be seen to embrace the historic facts of the situation.

The site, which features information boards, a herb garden, water feature and the stone circle, provides historic information in a setting that encourages spiritual contemplation. Approval for these plans was granted on 3 November 1999. It is hoped that future visitors to the airbase will have the opportunity to view iconic monuments of the Cold War, alongside this commemorative memorial.

Chapter 8
Strait Street

With Emily Morrissey

Strait Street (Valletta, Malta) is full of contradictions, and the deeper one digs, metaphorically and literally through the dust and rubble, the more obvious this becomes. Malta identifies strongly with the Catholic faith (e.g. Boissevain 1993), yet at the heart of the capital city and World Heritage Site (Valletta) is a street – Strait Street – where vice, crime and sex have been among its defining characteristics for much of the nineteenth and twentieth centuries.

Following Malta's independence in 1964, and the steep decline in the numbers of British and American servicemen on the island, Strait Street effectively closed down. But even then such stigma existed as to ensure Strait Street remained empty and seemingly unloved for some 40 years. Marks and Spencer, on opening an annex in Strait Street in 2003, 'protected' its customers by designing a bridge that could bring them over the street rather than along it. But attitudes seem to be changing, and the fabric of the street, its former bars and music halls, and the signs and advertisements, appear to be driving this process through the memories they evoke. Yet aspects of the taboo remain. For example, our fieldwork suggests that sex isn't much talked about in Malta. Yet those that have lived and worked in Strait Street appear genuinely enthused and encouraged by our interest in constructing an alternative view to the conventional histories of Valletta, and in particular by our focus on this extraordinary street – Malta's 'street of shame' (Saxon 1965). Our interest as archaeologists, recording the fabric of the place, the buildings and the objects and traces they contain, coincident with the studies of the historian Victor Scerri (below) and oral historical studies conducted through the Mediterranean Voices programme (e.g. Casha and Radmilli 2005) are giving Strait Street a respectability and a place in Valletta's colourful history that we believe it deserves.

This is first and foremost an archaeological study, interpreting the place, its meaning and significance, primarily and predominantly through its material remains. We are researching a subject for which we think the material remains can provide new insight, a place where the cultural, political and religious contradictions are acute, and a time within living memory for which many questions remain unanswered and – for some at least – taboo (for another example of historical and contemporary archaeology addressing issues surrounding sex and prostitution, see

J. Schofield, *Aftermath: Readings in the Archaeology of Recent Conflict*,
DOI: 10.1007/978-0-387-88521-6_9, © Springer Science + Business Media, LLC 2009

Gilfoyle 2005). The chapter is also about heritage, and how heritage – broadly defined in a socially inclusive and holistic way – can begin to give credence and relevance to that which is often considered unacceptable (see Schofield and Morrissey 2007 for a broader perspective on this issue).

Culture and politics form the backdrop to our study. In 1975, Geoff Dench discussed aspects of this in a review of Maltese crime, a review that is worth briefly revisiting here. Dench traced Maltese crime back ultimately to the central dilemma in colonial Maltese society – on the one hand, a strict set of moral principles on sexual matters; and on the other, the reality of prostitution to meet the needs of the large garrison stationed there. Because sexual subjection was so offensive to religious sentiments, Dench said, it had been unmentionable, and that – combined with the intractability of the problem – rendered all sexual issues even more highly charged than usual (1975: 107). Dench goes on to suggest that it may be because of their impotence in this matter that the Maltese came to demand such uncompromising moral standards. 'Social reality was so humiliating that they have collectively turned their backs on it in favour of the contemplation of an ideal and unattainable state of purity' (ibid.).

He even suggests that Wilhelm Reich's propositions on sexual repression and religiosity may find support among the bars and brothels of Strait Street (ibid.; Reich 1972).

Valletta's early history provides further context to our research. The city was planned and built by the Knights of the Order of St. John in the sixteenth century. It was arranged as a grid with fortifications encircling the city (these being the main justification for its inscription as a World Heritage Site). Some streets were wide to ease the movement of troops and for public display and ceremony. Others were narrower residential streets; some with workshops and shops. A few were narrower still, being little more than alleyways. Strait Street is just such a street, running from one end of the city to the other. Traditionally in urban design width reflected status, and Strait Street (along with a few other alleyways at the city's bottom end, close to Fort St. Elmo) came at the foot of the hierarchy, characterised by poorer residency, poverty, ill-health and (almost inevitably) petty crime. A hospital for 'incurable' women was located at this bottom end of the town, for example, established by at least 1625 (Cassar 1964: 70), while a health map of the city, dated to the 1850s, showed the instances of 1813 plague and deaths by household – this area fared worst, by some considerable margin.

Prostitution was prevalent in Malta certainly from 1530 when the knights took possession. By 1551, for example, the town of Birgu, where the knights first set-tled, was notorious for the many Greek, Italian, Spanish, Moorish and Maltese courtesans that lived there (Dissoulavy 1940, cited in Cassar 1964: 224). In plan-ning Valletta in 1566, it was intended to reserve a portion of it – the 'collachio' – for the sole use of the knights, and to confine Maltese prostitutes to 'some remote part' of the city. Later the knights introduced the periodical examination of women, a practice maintained at least to the middle of the nineteenth century. Until May 1832, women were examined at Strada Tramontana (South Street), Valletta. As Cassar says, however, '[o]wing to the "indecencies" which occurred on the days that the Police Physicians visited these women, it was resolved to transfer the clinic to a room under the venereal wards of the Women's Hospital and to place a sentry round this hospital to disperse the individuals who collected there ...' (1964: 228).

Between the 1890s and 1930s, the Maltese became more sensitive to their national image and the effects vice might have upon it. Consequently, the numbers of women recording themselves – or allowing themselves to be recorded – as prostitutes, declined: the numbers of *femmes declassées* in national census returns for example dropped from 136 in 1881, to none in 1930. In later censuses too, none were mentioned. This does not however represent a drop in business; merely the fact that government statisticians and others were aware of the threat to Malta's national honour. In police records, for example, the number of offences against regulations respecting public prostitutes rose from 268 in 1912–13, to 3,307 in 1923–24 (Dench 1975: 112–114).

Strait Street today is effectively two streets, each part with distinctive characteristics (Fig. 26). First, the upper (south) end, close to the city gate, where businesses and houses predominate – this end is clean and busy. Second, the lower (north) end, where the traces of former bars and music halls are more numerous, and where little except these abandoned business ventures remains – here cats prowl, birds feed off discarded food and the sounds are of music and conversation from the flats above. Few people walk in the lower end of Strait Street and the sense of abandonment is acute. Our initial interest in Strait Street was precisely because of this sense of emptiness, and of closure (in the physical, archaeological sense), and denial (in terms of memory and recollection). What material evidence did these buildings contain for their former use? What stored memories might their investigation 'unlock'? What significance did these hidden spaces hold to local people who supposedly would not be interested in our attempt to rake up a 'shameful' past, and its connection with prostitution and vice extending back 500 years? Faith, sex and heritage came together here in a fascinating weave that we intended to begin to unravel. The study could conceivably have taken place in any number of places (Gibralter for instance, or Singapore's notorious 'Boogie Street'), but Malta seemed more suited than most to the type of study we envisaged. This was largely because of the scale and speed of abandonment (most similar streets have either continued in use, or have been redeveloped), but also because of current proposals to regenerate this part of Valletta, and the likelihood that much of the fabric of the street would be lost without record if left only to local planning controls.

Two key components of Strait Street – bars and prostitution – were intriguingly and inextricably linked, however innocent the bars themselves may have been. As Dench points out, with regulations on vice, it simply would not be possible 'to restrict the soliciting of prostitutes in public and to hold back the boom in bars and floor shows' (1975: 116 – original emphasis) that were central to servicing the large garrison present in Malta. This was a connection we particularly wanted to explore and it is for this reason that our archaeological enquiry began in the abandoned bars at the lower end of the street.

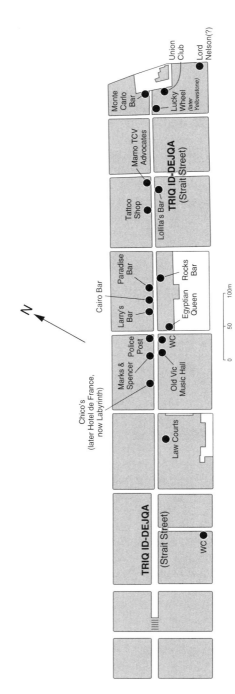

Fig. 26 Map of Strait Street showing sites mentioned in text. *Drawing*: Eddie Lyons

Progress in Strait Street

Our examination of Strait Street involves three stages of work: First, being seen in and around the street; second, the conduct of informal interviews 'in place'; and third, seeing and recording as many of the places and objects we encountered as possible. Being seen was an essential first step. We needed to build trust and confidence amongst those that lived and worked (or just hung out) in this part of Valletta. Some students working here in the past had been 'roughed up', and even quite recently there are reports of people being propositioned. We walked up Strait Street many times, spending at least 4–5 h each day doing so. We met people this way, and the contacts we made gradually produced the network of informants, owners and occupiers that we needed. Our interviews followed the advice of social anthropologists such as Anderson (2004) who adopted 'bimbling' as a methodology for investigating sites of environmental protest. For Anderson (ibid.: 257), bimbling amounts to walking or wandering as a means to providing the ideological space necessary to re-experience people's connections with landscape. Given the people we were meeting (one-time barmaids, cross-dressers, cabaret artists and – potentially at least – former petty criminals and prostitutes), we sensed that structured and formalised interviews – including questionnaires – were to be avoided, and that bimbling might be a sensible alternative, given the strong associations with place amongst most people we'd met, and the centrality of that association to our project. Through bimbling we were told about Strait Street, our contacts providing details additional to but occasionally repeating those gathered by our friend and local historian Victor Scerri, whose interviews are published periodically in the left-wing Maltese-language newspaper *it-TORĊA* (similar articles were later to appear – in English – in the *Times of Malta*). But most significantly, bimbling – combined with Victor's knowledge and influence – literally opened doors for us, and we gained access to ten of the 60 or so bars that originally lined the lower end of Strait Street. We had been told we'd never get into any of them, so this seemed like a result.

Results

This short summary is focused on the physical places that we encountered, and the insight they provided to Strait Street. The bars we visited surprised us, not so much for what we found, as the owners' attachment to the places, their use of them in the intervening years, and their attitudes to our own interest and enthusiasm. Here we provide four examples to illustrate this point.

At the lower end of Strait Street, on our first day of fieldwork, we were trying to decipher graffiti and lettering (highlighting the Cambridge Music and Dancing Hall) in bright morning sunlight. On the opposite side of the street, a door was open. Inside it was dark, and noisy – someone was making something, hammering and using a blow-torch. As we entered this gloomy space, a figure stopped working and removed his protective face mask. We walked towards him, noticing as we did so a large mirror on the far wall, and then the bar beneath it. He introduced himself as the son of the former owner of this, the Monte Carlo, bar. His father had opened the bar, built the furniture and run the business for many years, until the decline in trade necessitated closure sometime in the late 1960s or 70s. In honour of his father, he kept as much as was practical, and with obvious pride he showed us around. The beautifully crafted art-deco bar, now used to store the tools of his trade; the cash-till; the beading on the walls; the bell by the front door – hidden by a flap – with which, out of hours, the doorman would warn those inside that police were approaching. Here was a man who saw no shame in Strait Street. The photograph he gave us, of him standing at the bar with his mother (Fig. 27), is clear evidence of that. He even gave us three *landa*, the tokens with which barmaids were paid, one landa for each drink they persuaded servicemen to buy.

Further up the street is a typical example of one of Strait Street's smaller more intimate spaces – formerly the Rocks Bar. The owners still occupy the building, many years after the bar closed, using it almost as an urban beach-hut, with cooking facilities, a religious shrine and a loo. Here too there is pride in ownership. On entering, we first noticed the wonderfully intricate ceiling decoration: a wooden trellis supporting a tangle of artificial flowers, now sadly removed. The loo has long gone, but the graffiti carved on its walls remain, a deliberate choice by the owner. And in their storeroom at the back, the real treasures are kept: framed and signed photographs of the ships whose sailors returned to Rocks Bar on their visits to Malta; photographs of their friends in the bar; and a flag from an American ship, carefully folded, pristine and opened with great ceremony for our benefit.

The third and fourth examples are owned by young brothers that inherited their bars from their father, who owned several in the street. One is managed as a private club, and still has a bar – the layout and the decoration have changed, but functionally there is continuity here. The second is another small bar, similar to others we saw, that may simply have been a sitting room, converted as a commercial venture to cash in on lucrative business opportunities. Here – in what was Lollita's Bar – the second brother has a recording studio, making good use of the original sound proofing. He keeps the bar sign in his store room at the back.

In fact, nowhere in Strait Street, in interviews, meetings and particularly in visiting the former bars (and one music hall) did we witness any sense of shame. But away from the Street, things were different. We were told of reputable businessmen – 'particularly jewellers' – who made their money in Strait Street before moving into retail. We were then introduced to a jeweller who seemed over-keen to impress upon us just how clean and innocent Strait Street was, persuading us that our account of it should be 'respectful'. He posed for a photograph holding a religious icon. We were also warned off our project by one correspondent, responding to our call for information in *The Times of Malta*.

Former sailors often revisit Strait Street. On several occasions we met former (British) sailors, with wives and girlfriends, trying to locate bars using the few bar names that remain above the doors as the basis for reflection and reconstruction.

Fig. 27 Joseph Buttegieg with his mother in the Monte Carlo Bar. *Photographer* unknown

Here too are people that wish to remember Strait Street as it was. A tattoo artist who works from his shop in Strait Street gave an interesting dimension to this sense of reconnection: his father and his grandfather were both tattoo artists, working from the same shop, and today former sailors return to have their tattoos refreshed, by the son or grandson of the original artist.

The bar signs that remain are important markers – psycho-geographically they are central; they are the key reference points that provide spatial orientation for those that once knew Strait Street and continue to visit it today (Fig. 28). But they also present a dilemma: should the signs remain where they are, in situ; or should their owners – motivated by pride and attachment – be allowed to 'rescue' their signs for safe-keeping, and as family heirlooms, removing them from view? Is it legitimate for the original owner to reclaim 'his' sign? What if ownership has changed, and the new owners have no connection with the bar's former use? Should the new owner be obliged to retain and preserve the sign, or can he remove it at will? Whatever the answer (and it may be that like 48 'historic' timber shop fronts in Valletta, some of these bar frontages should be given legal protection, cf. Magro Conti and Darmanin 2003), the signs and the bar frontages do matter both as geographical markers, and as evocative and often stunningly intricate artefacts. Jon Mitchell (2002: 60) describes being shown around Valletta by his tutor: 'As we walked up the gently sloping street, he told me to write down the names of the various establishments – The Cotton Club, Cape Town, Garden of Eden – all names he associated with his childhood, with home and with the erstwhile vibrancy of this now defunct red light area of a declining city.'

Plastic flowers, a secret bell, sound-proofing and the elaborate and colourful bar signs above the battered doors together contribute to the material culture of Strait Street as we have defined it thus far. Much more remains to be uncovered behind many other doors, some of which haven't been unlocked for decades. But there are other aspects to Strait Street's material culture, and again they contradict and challenge the street's shameful associations.

Fig. 28 Several bar signs remain in Strait Street, including this one for the Smiling Prince. *Photo*: author

Representations

Close to Strait Street, one block away, is a building that may once have been a brothel. Indeed it may have been designed and built specifically for that purpose. It is an old tenement building with ten rooms on five floors, all the same and each with a tiny sink and a bucket propped underneath; at the back of each room was a loo. This was the building chosen to house *Cityspaces*, an exhibition of work by 11 of Malta's leading contemporary artists. Raphael Vella organised the exhibition, and said of the building: 'There was nothing much to say about the place, except that it was ugly and rather depressing. But the fact that it might have been a brothel made it much more interesting.'

Mark Mangion's response was typically in tune with his surroundings. Mangion placed a laptop in 'his' room, screening disturbing images, surrounded by photographs of a couple having sex in positions detailed in the *Kama Sutra*. Another artist filled the walls with pencilled diary entries recording the comings and goings of servicemen to a nearby brothel. In other rooms religion was the dominant theme, including (by Pierre Portelli) an armchair covered in wax representing Maltese churches ablaze with candles (information from www.studio-international.co.uk/painting/malta. htm – accessed July 2006).

Musically, Strait Street has a rich heritage. Before the introduction of the juke box, probably in the Second World War, live music was predominant, and this continued at some bars into the post-war period. Malta has a strong musical (and particularly jazz) tradition, and part of that is due to the venues and opportunities Strait Street provided to people like Joe (Il Pusé) Curmi, one of the musicians featured on a collection of historic (and recent re-) recordings from the bars and music halls of Strait Street (Galea 2004), an extract of which can be heard on our project Website: www.straitstreet.com.

Most recently, Strait Street was interpreted through contemporary dance. The significance here isn't so much the content or quality of the performance, but the subject matter and its status as another example from the arts of a willingness to confront this aspect of Malta's past. As Felix Busuttil, the Director of YADA (the Young Actors and Dancers Association), said to me in a text message of their performance of a work recalling Strait Street: '*Strait Street* is not a whimsical or thorough events (sic) of facts that have happened in the street we, Maltese, look upon with shame, curiosity and try and keep closeted. This is a journey into the fantastic world of the body … its passion, its love, its emotions, its senselessness, its sensefulness, its splendour, its downfall …'

We come full circle therefore, from sex – historically the very cause of Strait Street's 'shame', to the street's physicality – the sense of place and pride felt by those most closely involved with it, to creating a record, an alternative archaeology of it – interpreting it through the conventions of archaeology as well as through contemporary art and dance … and back again to sex. But within this circular progression, a gradual transition occurs from shame and negativity to pride and a far more positive approach to the obvious difficulties of Strait Street – a celebration almost. Perhaps it

is the point at which material culture is drawn into the equation, and with it a willingness to work with and interpret the physical place, that the very point of the transition
is reached (Dolff-Bonekaemper's point about discord value; the benefit of places that
enable a debate that might not otherwise occur, nd). Sex at this point becomes central
to the dialogue, and a necessary subject, if people – community groups, planners,
politicians – are to understand the place sufficiently to deal effectively and appropriately with its future management. In these terms, sex becomes an acceptable part
of the narrative, and the street a place that some at least can take pride in. Such is the
contribution contemporary archaeology can make to cultural political studies, and to
heritage. It is our hope that this alternative archaeology of Strait Street will be at least
acknowledged and perhaps even accommodated in plans for its future.

Section 3
Landscapes of Events

My choice of section heading here can be taken in one of at least three ways. First is the degree to which key battles or military episodes of the twentieth century have defined the epoch, and the necessity to consider their cultural impact at a broad landscape scale, a global scale even and in the context of, for example, ideas about industrialised warfare, 'Total War' and Giddens' (1999) thoughts on modernity. Second is the more literal study of landscape archaeology, in this case how topography/terrain influenced the course of a battle, for example; or why airfields were sited where they were. To study a particular bunker in close detail is of course a legitimate archaeological exercise, just as the close investigation of a prehistoric burial mound or medieval castle would be. But for both of these subjects, and for recent conflict especially, a regional scale of enquiry is required to make real sense of the event and the traces and legacies that remain. Then, third, there is the slightly more complex interpretation, closely matching Paul Virilio's interest (or obsession) with militarism and with time over space. To quote Tschumi (2000: viii):

> In Virilio's global temporal space, landscapes become a random network of pure trajectories whose occasional collisions suggest a possible topography: here is a peak, there an abyss. … Each collision is an event relayed by media [meaning that for Virilio] events are less here than now. His definition of the event is less in space than in time. Virilio's thesis may be simply that time has finally overcome space as our main mode of perception.

We should add to this the view of landscape, now widely used and accepted, as 'an area, as perceived by people, whose character is a result of the action and interaction of natural and/or human factors' (European Landscape Convention [ELC] 2000).

In this section, my landscapes of events involve an element of all of these: landscapes in the spatial, physical and perceptual sense; landscapes at local and global scale; and landscapes in time as well as in space. Certainly the emphasis of the book does begin to shift at this point, from being about place and the social value of physical remains in terms of memory and place attachment, to something more detached, broader and thematic. The distinction is similar to what Porteous (1996) refers to as 'autocentric' ('hot', physical, more 'actual' and close range) and the allocentric ('cool', intellectual, more detached and distanced) perspectives through which different groups of people experience place. If Sect 2 saw my more intimate side, engaging with place in a very personal way, in terms of both my own experiences and those of others whom I know well, Sect. 3 shows my professional capacity, as a heritage practitioner, working within a state heritage organisation and taking seriously my responsibilities in heritage protection and archaeological resource management. Only in Chap. 12 does the 'autocentric me' start to reappear, in good time for Sect. 4.

But with the temporality of landscape (Ingold 1986) comes distance, a remoteness that contrasts the close proximity and place attachment evident in earlier sections.

The chapters in this section are presented in chronological order. Chapter 9 deals with material culture on the home front during the First World War. With so much emphasis on the material remains and monuments of the Front itself, what remains at home has been very largely ignored by academic researchers and heritage professionals. Saunders describes the contradiction thus:

> On the Western Front, the Great War brought cataclysmic disorder to large areas of northern France and Belgium. Yet, this destruction of land and life created new landscapes infused with new meanings – a reordering of existence whose memories and associations came into conflict with other realities after 1918, and continue to do so, at an accelerating pace today. Associated in time, but not in space, were facsimile landscapes (i.e. training grounds), ambiguous 'spaces', where men practiced 'safe killing' under the illusion that Salisbury Plain was in fact the Somme. (2004a: 8).

This chapter originated as a short presentation at a conference on 'Material Culture, Memory and the First World War' at the Imperial War Museum (London), on 8 September 2001. The date will probably have some resonance, as the conference preceded an event which appears to have created a whole new experience of globalised conflict. As the organiser and editor said: 'In the view of many, the post-"9/11" world is a changed place, and attitudes that were held before it now seem almost as naïve and distant as those which were held in 1914 must have done to those of 1939' (2004b: 1). This chapter may focus on the relative safety of the home front, but there was no escaping the reality of what lay in store. In the trenches on the Western Front and the Home Front, just as at New York's Ground Zero, 'there is a terrible presence of absence' (ibid.). For this reason alone, for charting this course and seeing the historical perspective for contemporary events, these earlier sites of conflict should hold value for us all.

A preservation ethic and issues of national importance and cultural value are all central to Chap. 9. This chapter was published after I attended (in a non-speaking capacity) a national conference on The Battle of Britain, hosted in 1999 by the University of Edinburgh's Centre for Second World War Studies. The conference was popular and the coverage of subject matter was impressively diverse. A highlight was a session comprising two presentations by former pilots: Hans-Ekkehard Bob and Wallace Cunningham who entered into a friendly 'our planes were better than yours' type exchange. A photograph of the two pilots shaking hands appeared in the national press. To my mind the only thing missing from this collection of views and perspectives was conservation – commemoration was covered, but not conservation. As English Heritage had recently completed a national study of aviation sites from the Second World War, and was grappling with the issue of historic interest in addition to the usual architectural criteria, this seemed an opportune moment to publicise our emerging thematic study of this period.

The book is worth seeing in full, for its diversity and the close attention to detail in some contributions (Addison and Crang 2000). A review by Nicholas Lezard in *the Guardian* Newspaper (where it was 'Pick of the Week') noted four outstanding contributions: the two pilots', Nigel Rose's touching, indiscreet letters to his parents while with 602 Squadron, and Owen Dudley Edwards' 'mad but enthusiastic piece on "The Battle of Britain and Children's Literature"'. Lezard enjoyed the book, for 'it poses questions about the morality of war, the existence of heroism, the reliability of memory … it treats the subject honestly and with justice'.

At around the time English Heritage was investigating and assessing military aviation sites, attention also turned to another momentous event that was attracting public attention: D-Day. As with the First World War, the significant action took place overseas, though the preparations and planning were from home. Two events sparked this interest: First was media coverage of the anniversary celebrations at the Normandy beaches in 1994, with images of the US President and other dignitaries inspecting ruined bunkers amongst the dunes. This prompted letters asking how English Heritage was dealing with Second World War heritage in general, and our D-Day preparatory sites in particular. A study of primary sources was the result at this stage. But later, in 1998, one of our more spectacular D-Day monuments in England, the embarkation 'hards' at Torquay, came under threat from redevelopment. There was a hue and cry about this, and a national study of surviving remains was required to give this particular site context. During some wonderfully dry and hot weather in summer 1999 site visits were made, and national importance defined and ascribed to a selection of surviving sites, mostly in the south west of England. It is that process and the results of the endeavour that are reproduced here.

In Chap. 12, we move to the Cold War, another subject that English Heritage has given close attention over the past 10 years (Cocroft and Thomas 2003). In this essay I investigate the extensive and complex structures left to us by the Cold War. As the editor said in the journal in which the essay was originally published,

> Although he would not approve, William Morris's words seem particularly appropriate: 'but every change, whatever history it destroyed, left history in the gap, and was alive with the spirit of the deeds done midst its fashioning' (ASCHB Transactions 2001: 2).

The essay may seem a strange one. Written to be read, and to be visually and aurally challenging, the essay was presented to an audience of conservation architects in a room at the Tower of London. I showed a deliberately diverse selection of striking and disturbing images, and played some music composed on the theme of Cold War surveillance. I am sure they considered it a very strange presentation indeed!

Virilio's fascination with speed and technology lurks in the background throughout this section. The subtext examines how in just 80 years, from the Somme to the Second Gulf War, we went from seeing newsreel footage some weeks after the event, and letters from loved ones received months after they were written, to a world where we witness war in real time on our television sets; where combatants stay in touch with their families by mobile phone, skype and email.

And how the world has changed from slugging it out in the mud, to the capability in developed countries of ensuring annihilation with precision bombing that can take out very precise targets from afar. Even now, writing this as a white, middle class, middle aged man, it is hard to recall the climate of the Cold War at all, even though I was in my present job before the Berlin Wall came down. Sometimes it seems as distant, as remote as those prehistoric and medieval landscapes that are traditionally the focus of archaeological attention.

These chapters are reproduced by kind permission of the original publishers. The original citations follow: Schofield, J. (2004), Aftermath: materiality on the home front, 1914–2001. In Saunders, N. (ed), *Matters of Conflict: Material Culture, Memory and the First World War*, pp. 192–206. London: Routledge; Lake, J. and Schofield, J. (2000), Conservation and the Battle of Britain. In Addison, P. and Crang, J.A. (eds), *The Burning Blue: A New History of the Battle of Britain*, pp. 229–242. London: Pimlico; Schofield, J. (2000), D-Day sites in England: an assessment. *Antiquity* 75(287): 77–83; Schofield, J. (2003), Conserving legacies of the Cold War: An excavation in five parts. *Transactions of the Association for Studies in the Conservation of Historic Buildings* 26: 37–45.

Chapter 9
The Home Front, 1914–18

For a war fought almost entirely on foreign soil, the 'Great War' has had a remarkable impact on British society, no section of which was immune from its repercussions. Manners and morals changed as a result of the War. The working class and the nation came closer together, though due more to what the ruling class now considered to constitute the nation rather than to a change of attitude among the working class (Bourne 1989: 227). And there were other more obvious and visible changes to British life. For example, smoking increased in popularity, and men and women smoked more publicly than before. Swearing became more socially acceptable. For women, hemlines shortened and hair styles became more practical; more *mannish* (ibid.: 235). More women worked and there was a change in the nature of the work that women undertook. More important still was women's belief in what they could do, and society's belief in what they might be required to do. The consequences of this remain with us (ibid.: 198).

Less obvious are the effects of militarisation, many aspects of which remain legible today as some of the more tangible traces of the War on the Home Front. First came the preparations for war: the armament and rearmament of coastal fortification and the construction of anti-invasion defences; and the sites and buildings concerned with explosives manufacture – the production of *matériel*, a characterizing feature of this first industrial war, and one clearly reflecting the gender divisions between actions at home and on the fields of battle (Saunders 2002). Second was the emphasis on military training, including the preparations for trench warfare, for gas, and the pioneering phase of aviation – a feature of both the immediately pre-war and war years. Third was the physical impact of the war itself; for example damage to coastal towns in north-east England from enemy craft at sea and in the air. Finally are the places of memory and commemoration created and maintained in the post-war years in the form of war memorials, museums and their associated landscaping and architecture which constitute a further dimension, a further layer to this materiality of the Great War. While clearly related, remembrance is a subject that has received recent critical attention (e.g. Tarlow 1997; Winter 1995; Black 2004). For that reason, and as an archaeology more of remembrance than of the War itself, this subject is deliberately excluded from this assessment.

J. Schofield, *Aftermath: Readings in the Archaeology of Recent Conflict*,
DOI: 10.1007/978-0-387-88521-6_10, © Springer Science+Business Media, LLC 2009

This chapter will examine this archaeology of the First World War 1914–18, outlining and assessing what survives in England, and what it contributes to our understanding of the Home Front. A final section will contrast it with materiality across the English Channel, where the battlefields themselves will inevitably generate a more intense emotional response in visitors than training areas and coast batteries can possibly achieve at home. However, the point will be made that both records form significant components of this 'total war' and both therefore merit recognition albeit for a slightly different combination of reasons.

Remembering

The material culture of the Great War had relevance almost from the moment of its creation, initially for reasons of remembrance, cultural tourism and understanding; later (and additionally) as a means to interpreting past events in a landscape no longer so easily read as a battlefield. But battlefield tourism has been there from the start. As the introduction to the republished *Michelin Guide to the Somme* points out, while newspapers of the day did not tend to publish photographs from the war zone, other journals did, 'and from magazines like *The War Illustrated*, *The War Record*, *The War Budget* and *The Illustrated London News*, the public gained some impression of what the battlefields were like' (Peacock 1994). Films such as *Britain Prepared*, *The Battle of the Somme* and *The Battle of Arras* also gave an impression of conditions at the Western Front. But it was all sanitised, and for those left at home there was an intrinsic sense of curiosity to see what it was really like, as soon as the opportunity arose.

It was against this background that the *Michelin Guides* were published, alongside others such as Capt. Atherton Fleming's *How to See the Battlefields* (1919). These and other comparable guides typically show plans, portraits of key figures, general battlefield scenes, cemeteries, damage to cultural property, and what we would now describe as monuments of war: bunkers, observation posts, trench systems and so on. People touring the battlefields wanted to see these structures but more especially they wanted to witness for themselves the conditions of the front; it was the most effective way to feel the experience of war in what was then a silent place with a clear and tangible sense of sanctity, facilitating quiet reminiscence (King 1998: 229). They visited these places perhaps also for therapeutic reasons, for reasons of guilt (amongst those that stayed at home), to help comprehend the scale of the conflict, but nearly always for reasons of remembrance and mourning. Whatever the motivation, these structures and places played an important role for those left to rebuild society and their own lives in the immediate post-war years.

As a consequence, and for the related reason that unexploded ordnance typically hampers any clear-up operation (Webster 1997), these historic resources have survived comparatively well, and now once again play a significant role in cultural tourism in this region. Furthermore, now the monuments themselves also play a role in commemorating the war dead and remembering the fallen: some cemeteries and memorials now incorporate bunkers or concrete fragments in their design, while the presentation of other sites (like Vimy Ridge with its trench systems and shell holes) is periodically re-appraised to meet changing standards, perceptions and expectations (Cave 2000).

By contrast, what survives away from the front has only accumulated cultural values much more recently, and within the context of a developing interest in the archaeology of the recent and contemporary pasts (Graves Brown 2000; Buchli and Lucas 2001), in military archaeology (Dobinson et al. 1997; English Heritage 1998; Schofield et al. 2002), and with the growth of popular interest in military history and its spin-offs in publishing, the cinema and television. In England, numerous related projects and studies have combined to provide a record of First World War activity (some forming part of wider surveys commissioned by English Heritage to inform conservation decision-making), and it is a rapid review of these initiatives that forms the basis of this chapter. At the end, I will return to the motivations for preserving components of this materiality on the Home Front.

Cultural Resources

Defences

The Riddle of the Sands, Childers' (1903) fictional account of the preparations for an enemy invasion of Britain across the North Sea, was ahead of its time, but not by as much as some might imagine. Britain was prepared for invasion during the First World War. As the German army advanced through Belgium to Ostend, it was estimated that invasion of Britain could be undertaken by a force comprising 70,000 men carried in barges. It was furthermore realised that naval intervention involved a 24–28-day delay, thus requiring some further anti-invasion measures to be put in place (Wills 1985). These included a series of stop lines comprising fieldworks and pillboxes, designed to prevent or slow an enemy advance. An earlier line of London Defence Positions (based on a number of mobilisation centres) was brought back into use, as was a defence line at Chatham (Smith 1985) where entrenchments with pillboxes were built between Maidstone and the river Swale (Kent). A further three lines were constructed to the north and east of London. Pillboxes were also built along the east coast, some facing inland to prevent ports from an overland attack. In Suffolk and Norfolk, these early pillboxes were circular in plan, contrasting with those in Kent, which were hexagonal, similar in form to later Second World War examples. To give an idea of scale, 300,000 troops were deployed on the east coast in winter 1914 to man these defences (Saunders 1989: 213).

Today 11 of the 13 mobilisation centres survive, with some of the buildings remaining in use for accommodation and storage. Those at North Weald (Essex), Alderstead (Surrey) and Farningham (Kent) are among the best preserved. Pillboxes have been recorded as part of the Defence of Britain Project, a now-completed review of anti-invasion defences in the UK, revealing that some 50 First World War examples survive, about half of which remain in good condition.

Britain's coastal defences were well prepared at the outbreak of war, owing both to the close attention paid to home defence and the realisation of the German naval threat over the previous ten years. However, on only one occasion, 16 December 1914, coastal defences were required to fight off German warships. This was at Hartlepool following earlier attacks on Great Yarmouth and Gorleston. Saunders (1989) describes the event (see also Dobinson 1999a: 118–119):

> Hartlepool was defended by Heugh Battery with two six-inch guns, and Lighthouse Battery with just one. The first shell from the [battle cruiser] *Seydlitz* fell between the two batteries cutting all the fire-commanders telephones. In spite of many shells falling close to the batteries there were only four fatal casualties among the gunners, though 112 civilians were killed in the town, and much damage done to its buildings and docks. The coast defence guns, despite the initial damage, hotly returned the enemy fire. It was held to be a creditable performance by a severely under-gunned coast artillery unit, and the principal members of the batteries were decorated (Saunders 1989: 209).

But although Britain was in a state of readiness, substantial additional works during the War were needed especially to east coast batteries (indeed, as Dobinson (1999a: 46) has observed, for coast artillery the period 1914–18 was more a building programme than a war). A heavy battery (Brackenbury Battery) was constructed at Felixstowe (Suffolk) to provide added protection for the Harwich approaches, while defences were also placed in the Humber and Tyne, at Plymouth and in the Bristol Channel. Batteries were also built to flank the Solent boom defences. Of the coastal batteries in use during the First World War, 35 were newly opened in the period 1900–14 and a further 23 between 1914 and 1921.

Using documentary sources, English Heritage has completed research into coast artillery 1900–56 (Dobinson 1999a), with a subsequent assessment of what survives (Schofield 2002a: 277–279). Of the 286 twentieth-century coast artillery sites, 35 are well preserved; a further 129 remain in some form. Of those that no longer survive the majority are Second World War Emergency Batteries. Most that were newly opened in the period 1900–21 have some surviving remains, as do those of earlier date.

Another significant group of sites was anti-aircraft and airship defences, in the form of artillery or gun sites with their associated searchlight positions. These guns were positioned to provide defence against aerial attack by Zeppelins (causing in all 557 fatalities and £1.5 million worth of damage during the First World War) and later, Gotha bombers (one raid on London in 1917 killed 162 and wounded 432 people). A review of these sites, again based on documentary sources, has been undertaken (Dobinson 1996a: 11–47, 2001: 3–58). Sources describe how the majority of First World War anti-aircraft guns in Britain were fixed, established at permanent sites, many of which were purpose-built. They also show how the emphasis shifted through the war, from defending military targets early on, to the protection of civilian targets by 1916 (Dobinson 1996a: 11). The total number of these sites in England is currently documented as 376, at which few traces are likely to survive.

In terms of passive air defence, experiments with acoustic detection began during the First World War, with the result that concrete sound-detecting acoustic dishes were built in at least eight locations around England's south and east coasts (Dobinson 1999b: 8–12). These could, in theory, pick up the sound of an approaching aircraft at ranges of 13–24 km. They proved unreliable however, and this technology in any case was soon overtaken by experiments with radar. Some examples of these earliest sound mirrors survive on the north-east coast. In addition, there were radio-telegraph stations for ship to shore communications. These were located around the British coastline in 1915 though most had been closed down and removed by 1920 (Sockett 1991). Where airships or aircraft did get through, they were often confused as to the location of their intended targets by primitive decoy sites, lighting arrangements designed to mimic the real target. These decoys were successful on occasion (Dobinson 2000a: 2–3), but were only ephemeral structures, and it is unlikely that anything of these First World War examples will survive.

The Production of War Matériel

Munitions production has been the subject of a recent and comprehensive overview (Cocroft 2000). As Cocroft states, only four days after the declaration of war – on 8 August 1914 – the first of the Defence of the Realm Acts (DORA) were passed, giving the government powers to acquire land for the prosecution of the war, and to control everything necessary to make munitions (ibid.: 155). This involved the supply of essential raw materials such as acetone for the production of cordite. A notable survival is at Holton Heath (Dorset), where the first purpose-built plant exists to exploit the Weizman process by which starch sources (in this case maize) were fermented directly to acetone. This plant survives as the footings for a large barn for storing maize, a cooker house for reducing it to mash, and six of the original eight fermentation vessels, later adapted to serve as air raid shelters in the Second World War.

Factories fell into three categories: propellants manufacture, high explosives manufacture and National Filling Factories. Of the first category, several existing sites were enlarged and continued in production (e.g. Waltham Abbey [Essex] and Cliffe [Kent], Fig. 29), while other new factories were created, as at Holton Heath and Gretna (Cumbria).

Today, much remains at Holton Heath, but little at Gretna which was largely demolished in the 1920s and later reoccupied by the army. In the case of high explosives manufacture, TNT was to become the standard filling for land shells, with lyddite important for naval shells. Early in 1915, there were ten TNT plants in operation, though by June this had risen to 16 (Cocroft 2000: 168). Purpose-built plants include Oldbury (West Midlands), while at Hackney Wick (London), the Phoenix Chemical Works were converted to TNT manufacture. Tetryl was also important, though as an intermediary explosive. A key site here is Waltham Abbey where tetryl production began in 1910, while another is at Holton Heath. Finally,

Fig. 29 A late nineteenth-century explosives works at Cliffe (Kent), still operating through the First World War. *Photo:* English Heritage (NMR 15033/25)

National Filling Factories at the outbreak of war were limited to the Royal Arsenal Woolwich (London) and factories at Lemington Point and Derwenthaugh near Newcastle. Other examples, including those for small components, cartridges and gas, came later.

Training

Although most battle training for the Western Front was done in France, military training was also undertaken on home soil and the traces of this survive as a cogent record of the preparations for combat. Although no synthesis is available (beyond an annotated list of military training establishments in England – Dobinson 2000b), recent work by English Heritage's survey teams on Dartmoor (Probert, personal communication), Salisbury Plain (McOmish et al. 2002) and Exmoor (Riley and Wilson-North 2001), and by others elsewhere (e.g. Welch 1997) has produced a record embracing all main phases of twentieth-century military training activity. In terms of First World War remains, the trench systems on Salisbury Plain, representing practice trenches dug from at least 1902, are the most complete and extensive to survive in the UK, amounting to one of the largest earthwork monuments on the Plain.

In some places, contemporary obstacles such as wire entanglements secured by screw pickets remain in place. As McOmish et al. (2002) describe, the trench systems were composed of three elements: Front Line, support and reserve – all of which were connected by a further series of communication trenches. In addition, shelters and smaller specialised trenches were constructed. An example is Perham Down, although now reduced by modern ploughing. This example covers over 100 ha and comprises at least three separate trench systems, illustrating 'the ebb and flow of warfare where successive firing lines were constructed as the battle progressed' (McOmish 2002: 142). A further example relating to trench warfare is the survival of concentric gas trenches on Porton Down, Wiltshire, now protected as a scheduled monument (see Chap. 1).

In addition to trench warfare, the emergence of the tank during the First World War has left its mark. Also on Salisbury Plain, on the edge of Shrewton Folly, an anti-tank range was constructed in 1916 comprising two parts: a firing line where artillery guns were deployed; and a target in the form of a hessian or canvas screen shaped like a tank and mounted on a trolley that was towed along a railway line at various speeds. The range covers 65 ha, much of which survives today.

Another dimension of military training, and one of social, historical as well as archaeological interest, is the badges and insignia cut into the chalk scarp at places like Fovant (Wiltshire) (Holyoak 2001). Nine of the original 19 badges remain visible at Fovant, a tradition started by troops stationed there during the First World War. The earliest badge (1916) is thought to be that of the London Rifle Brigade, the 5th Battalion of which is known to have been trained here between January and May of that year. Initially present during 1916–17, soldiers belonging to the Australian Imperial Force, Australia's expeditionary force, took over many of the camps around Fovant from October 1917 until after the Armistice. During this time, they cut the so-called Rising Sun, the General Service badge adopted by the Australian Commonwealth Military Forces from 1911 onwards. These badges and insignia are prominent and poignant features associated with a number of regiments or units either subsequently disbanded or whose members left Fovant to fight in some of the most bloody battles of the First World War (ibid.).

Aviation

Aviation has its origins in the immediate pre-war years of 1911–14, and with the onset of the First World War, the Royal Naval Air Service, and later the Royal Flying Corps, were given the air defence role. When the RFC took over this role in 1916, following air raids by long-range Gotha bombers and Zeppelins, Home Defence Stations were established in eastern England. In late 1915, training was decentralised to cope with the numbers of volunteers, and many new training aerodromes were established. From 1917, Reserve Stations for training pilots for the Western Front became the greatest airfield construction programme of the period and 301 sites were in occupation by 1918. Aerodromes in this early period were usually laid out as four groups of buildings: the officer's mess and quarters; regimental buildings; technical buildings (including hangars); and the women's hostel (Francis 1996: 12).

A review of aviation sites and buildings by English Heritage (Lake 2000) has demonstrated what survives from this early phase of aviation. The majority of buildings from this period were of temporary materials expected to last only for the duration, and were either cleared after 1918 (271 of the 301 sites) or have since decayed. Of those that remain, hangars survive on eight sites (including Calshot [Hampshire], Old Sarum and Yatesbury [both Wiltshire]). However, only one site (Old Sarum) has retained its suite of hangars (Fig. 30) and technical buildings fronting onto an airfield relatively unaffected by later development. This has now been designated as a conservation area.

Summary

There are other classes of monuments of course, such as hospitals (Richardson 1998: 98–100), internment camps and naval facilities that are not covered here. However, while it claims to be neither definitive nor comprehensive, this brief

Fig. 30 First World War hangars at Old Sarum (Wiltshire), still used for their original purpose. *Photo:* author

review of the materiality of the First World War on the Home Front does demonstrate the scale and diversity of material culture in the form of buildings and monuments, and the extent to which England's landscape was altered at this time for reasons of preparation, production and fortification. To give an overall impression of scale, it has been estimated that over one million acres were occupied or used for military purposes during the First World War. Some of these impacts on the landscape, and much of this material culture survives despite the fact that it represented an unfashionable and unconventional heritage until recently. But attitudes have changed, and just as the meaning of the Remembrance ritual has altered, in part due to a loss of emotional intensity and partly because those involved with its celebration have reglossed it (Tarlow 1997: 118), so the disposition towards these material records has changed. I now turn to the relevance of these First World War remains in contemporary society.

Not Forgetting

Have you forgotten yet?

S. Sassoon (1983)

Much has been written recently about the relevance of the recent past, and of its material records, whether merely *as a record* and to ensure significant past events are not forgotten, as cultural benefits now and in the future, for reasons of retribution, or for more personal and pyschotherapeutic reasons (Forty and Küchler 1999; The Ludlow Collective 2001). But how relevant are these considerations 'at home', when all the meaningful actions took place overseas, and at sites rightly recognised and treated as sacred?

A significant point here is what these sites on the Home Front represented within the broader context of the Great War. As Horn has said (2000), the First and Second World Wars both represent examples of 'total war', a defining component of which is that the division between military and civilian worlds is effaced and the Home Front is integrated into the practice of warfare. More specifically, 'the fighting men depended upon the merchant seamen and upon the work of non-combatants at home for the instruments of victory' (Woodward 1967: 453). So, if we consider the preparations and execution of warfare as a process, involving phases, events and social actions set within this wider socio-cultural and political context, and within a longer time-frame, than merely the battles themselves, then the activities on the Home Front, preparing and producing for war, and responding and reacting to it, form part of that process and are not separate from it. In other words, in total war, the spatial limits of warfare are extended from the battle zone to explicitly include the Home Front, making the classes of monument described in this chapter significant as 'reference points or landmarks to the totalitarian nature of war in space and myth' (Virilio and Lotringer 1997: 10), alongside battlefields and bunkers. To put this in more emotive terms, we might talk of such places as being symbolic of the sacrifice made by a lost generation; in more objective terms, we can talk of cultural assets and sites of national importance – places that represent the preparations for a war that was supposed to end all wars: the trench systems where soldiers practised before heading to the Western Front, the insignia they carved on chalk escarpments, and the coastal defences, including those which engaged the *Seydlitz*.

Either way, these are significant sites because, as Buchli and Lucas (2001: 80) have argued: 'From books to computers, from mementoes to war memorials, material culture shoulders the larger responsibility of our personal and collective memory. The corollary of this, of course, is that the decay or destruction of these objects brings forgetfulness.'

Along the line of the Western Front, battlefield tourism and the inherent danger in clearing explosives have meant that much material culture remains in the form of trench systems and dugouts, concrete emplacements and scattered *matériel*, to provide an experience for visitors. Some of these places speak for themselves; they have an atmosphere that is tangible and can easily be drawn out by reading the first-hand experiences of those who were there – Delville Wood on the Somme for example where, it is said, birds never sing, is a case in point:

In Delville Wood – in Delville Wood
The shattered trees are green with leaves,
And flowers bloom where cannons stood.
And rich the fields with golden sheaves –
Sleep soft ye dead, for God is good –
And peace has come to Delville Wood.

From A Soldier's Song, by Lt. Fred C. Cornell.

Of course it is the battles themselves that this material culture ultimately represents; the places where three quarters of a million British men died as a direct result of the War (recalling the total figure for the First World War of some 20 million deaths), and twice that number were disabled (Tarlow 1997: 110). It is therefore a material record strong on emotion, and one where interpretation has to be carefully managed to avoid the risk of trivialization. Care must be taken to ensure that interpretative motivations do not override the responsibility towards personal and collective memory; that a strong sense of the sacred is retained.

By contrast, what survives in England are those places where soldiers trained for the battles to come, where they prepared for an anticipated invasion and, in some cases, where civilians died as a result of air- or sea-raiding or in industrial accidents. These are also the places where *matériel* was produced – often by women – an industrial process that started at home and ended on battlefields like the Somme and Ypres. These monuments of the Home Front may not have the same emotional charge; they may not be sacred to the same extent, but they are cultural resources that tell of past times and events. Furthermore, they are all that survive of some aspects of the War (e.g. the role of women), thus giving them significance for interpretation, education and awareness.

As the growth of interest in military history shows, people want to know, and – we assume – will continue to want to know in the future. Retaining some sites and some material culture that represents military training and production, and coastal defences, will contribute to meeting that need. Some sites also have other more specific values. Aviation sites around Salisbury Plain for instance represent early experiments with military flying, while others – like Holton Heath – display the evidence for developments in industrial production, in this case biotechnology. This was an industrial war, like no other before it, and much of the evidence for that industrialisation survives on the Home Front, though the effects were seen most clearly and so terribly overseas.

Work is already underway to retain and present these sites. Waltham Abbey, for example, is now a visitor attraction, following a detailed survey of the site (described in Cocroft 2000). Many of the sites that survive have statutory protection. Another important issue is that the archaeology of modern warfare, specifically, here, of the First World War, is also an archaeology of us, reflecting our changing attitudes to conservation and to the need for preserving memories of past conflict in contemporary society (see Gilchrist 2003). It is only partly the case that once personal memories fade, the horror will be forgotten (Forty 1999: 6). The horror of the First World War can also be seen through engaging interpretative displays, with photographic images, first-hand accounts and – importantly – the actual places where conflict occurred. However, it can also be seen in the sheer scale of this war and how it enveloped all aspects of life at home and abroad – and here war memorials, with their lists of names, form one aspect, as do the many sites and areas in England where soldiers prepared for war, and where its *matériel* was produced.

Chapter 10
The Battle of Britain

With Jeremy Lake

When Richard Hillary, in the process of recovering from terrible injuries sustained as a fighter pilot during the Battle of Britain, sat down to write *The Last Enemy,* his thoughts turned from the story of men, machines and the camaraderie of arms to a growing sense of loss. His fictitious final chapter, an account of the death of a working-class woman in the Blitz, touched on the essential paradox of this battle (Hillary 1997: 174–176). This was that although the heroic 'few' and their machines were rightly eulogised as 'knights of the air' through art, music and film, the summer of 1940 witnessed the realisation that civilians were not immune to 'the new impersonality of warfare [which] turned killing and maiming into the remote consequence of pushing a button or moving a lever' (Hobsbawm 1995: 50). 100 civilian deaths in June air raids rose to 300 in July and 1,150 in August. The grim tally between 3 and 11 September was 1,211, including 976 in the London area (Hennessy 1992: 31; Pelling 1970). Whilst none of the civilians subjected to aerial bombing in 1940 could have foreseen the full scale of the horrors that were to lie in store for the civilian populations of the combatant countries of the Second World War, many would have been aware that the location of airfields and anti-aircraft sites in and around the great urban areas manifested a significant shift in the conduct of warfare.

The sites associated with the Battle of Britain reflect to varying degrees both the popular image of the battle, embodied in the heroism and sacrifice of the combatants, and the all-embracing nature of 'total war'. Famous place names such as Biggin Hill and Kenley, therefore, have a dual historical and cultural significance which in our view makes them worthy of conservation (Dobinson et al. 1997; English Heritage 1998).

Attitudes towards this heritage are, of course, inextricably linked to its complexity and associations. The bomb-shaped memorial erected in 1935 'in protest against war in the air' outside Sylvia Pankhurst's home at Redbridge (London) manifested a growing realisation that, despite the liberating potential of air technology, air power had turned civilians into military targets. The image in the mind's eye of the mass destruction of civilian populations, and of names such as Guernica, Dresden and Hiroshima, challenges our very notion of heritage. The protection of historic buildings, through listing under the *Town and Country Planning Act* of 1944 was, in fact, the outcome of a heightened awareness of the potential destruction of a nation's culture which aerial bombing brought in its wake (Saint 1996).

J. Schofield, *Aftermath: Readings in the Archaeology of Recent Conflict,*
DOI: 10.1007/978-0-387-88521-6_11, © Springer Science+Business Media, LLC 2009

We now recognise that the horror and 'hot emotions' associated with some sites of the more recent past should not preclude their preservation, and indeed that they embody society's duty of memory (Uzzell 1998). Illustrative of this is the identification for protection immediately after the Second World War of the concentration camp at the Camp du Struthof, at Nazweiler (Alsace), and the debate which followed the archaeological recovery in 1987 of the Gestapo headquarters' basement in Berlin, now conserved as a place where terror was planned and administered (Rurrup 1998). In a different sense, the ruins of Coventry Cathedral and the Atomic Bomb Dome at Hiroshima symbolise the devastating effects of war, the resilience of the communities directly affected by it, and the resurgence of economic, social and cultural fortunes.

In contrast to the popular interest in the *matériel* of twentieth-century conflict, including the military aircraft which attract thousands to museum sites, it has only recently been possible to consider more dispassionately the historical role and importance of the sites and buildings which made up Britain's defensive and operational infrastructure. The range and variety of structures involved is enormous, a direct reflection of the changing nature of external threats and the new and varied countermeasures built in response to them. These have had a profound effect on the landscape, from the construction of airfields, radar sites and anti-aircraft batteries, to the thousands of structures and earthworks associated with the anti-invasion defences erected throughout Britain in the summer of 1940. Prominent amongst sites opened to the public are the tunnels associated with the Dunkirk evacuations at Dover Castle, and the Imperial War Museum sites at Duxford airfield and the Cabinet War Rooms in Whitehall. Recognition of the importance of the last, constructed in 1938 as the 'central shelter for government and military strategists' during the Second World War, underlay the announcement in parliament in 1948 that it should be preserved as an historic site, although it was not until 1981 that the decision was taken to open it to the general public.

Local authorities and communities also recognise the value of such sites. Hence the involvement of Hackney Borough Council in the reopening of an air-raid shelter, built in Rossendale Street in 1938. Similarly, veterans and various private trusts and societies have participated in the conservation of the 1940 Emergency Coast Battery at Battery Gardens in Brixham (Devon) and the establishment of privately run museums at the Battle of Britain airfields of Hawkinge (Kent), Tangmere (West Sussex) and North Weald (Essex). Organisations such as the Fortress Study Group and the Airfield Research Group have stimulated the study and recording of twentieth-century military sites, and volunteers working for the Defence of Britain Project, initiated by the Fortress Study Group and the Council for British Archaeology in 1995, have undertaken a leading role in the systematic survey and recording of such sites (e.g. Foot 2006).

We have seen already that English Heritage is the government's statutory adviser on all matters concerning the conservation of the historic environment in England. Its statutory duties are, first, to secure the preservation of ancient monuments and historic buildings; second, to promote the preservation and enhance the character and appearance of conservation areas; and, third, to promote the public's enjoyment, and knowledge, of ancient monuments and historic buildings. In fact the terms 'ancient' and 'historic' are both something of a misnomer, as the scope of this legislation ranges from prehistory to the present day.

English Heritage's survey of twentieth-century military remains embraces all three of these statutory duties, but with more emphasis on the first and third: protection and understanding. In modern conservation practice, sustainability is the key, and this requires a sound understanding and characterisation of historic resources prior to making decisions about them. Before our national survey there had been no systematic review of sites, their typology, national distribution, vulnerability and rates of survival. Without such information, decisions about the preservation of individual structures could not be made on the basis of an informed judgement. In the following two sections we describe the outcomes of an English Heritage study of aviation sites, in terms of the types of monument and building relevant to the Battle of Britain, and we reflect on the strategies necessary for their conservation.

The Landscape of War

Much of the infrastructure of the RAF in the summer of 1940 had been deployed with a very different type of air war in mind. In contrast to the post-1933 Luftwaffe, the interwar RAF envisaged future wars as being fought from fixed and secure bases built in permanent materials. The first phase of this scheme, the principles of which were established by the Salisbury Committee of 1923 (Dobinson 1997: 25), took place under the guiding hand of General Sir Hugh Trenchard, who formed the RAF as the world's first independent strategic air force. His scheme involved the construction of offensive bomber bases in East Anglia and Oxfordshire, sited behind an 'aircraft fighting zone' some 15 miles deep and stretching round London from Duxford near Cambridge to Salisbury Plain. It was the creation of this zone which accounted for the rebuilding of several First World War bases around London: Biggin Hill (1914), Northolt (1915), North Weald (1916), Kenley (1917) and Tangmere (1918), and the refurbishment of the satellite stations at Hawkinge (1915) and Martlesham Heath (1917) (ibid.: 31, 34, 39, 41). Although political and financial factors prevented the completion of Trenchard's scheme, the collapse of the Geneva disarmament talks prompted the government to embark after 1934 on its largest interwar expansion of the RAF. This accounted for the construction of a number of new fighter airfields which were later to play an important role in the battle, including the 11 Group sector stations at Debden (1937) and Middle Wallop (April 1940), and Biggin Hill's satellite at West Malling (June 1940) (ibid.: 113–114).

The result was the construction of more than 50 new permanent stations between 1923 and 1939 and the extensive rebuilding of those retained after 1918, representing an unprecedented peacetime investment, far exceeding in real terms even the coastal fortifications built under Palmerston's administration in the 1860s. Standardised designs for every aspect of airfield operations proliferated, from bomb storage, simulated training, motor transport and storage to accommodation for all ranks, cinemas and barbers' shops (Francis 1996). The design of Trenchard's stations displayed a stark utilitarian architecture which – apart from the Garden City inspiration for station married quarters – owed much to the army background of the designers, who worked from the office of the Air Ministry's Directorate of Works and Buildings. But it was the need to provide for dispersal in the event of air attack which made airfield planning markedly different from the more condensed layouts of naval or army barracks. This is exemplified, for example, in Trenchard's requirement for the crescent, as opposed to previously linear arrangements of hangars, and the designs for officers' quarters, which dispersed the mess, recreation rooms and accommodation in order to obviate the risk of a single run of bombs destroying an entire complex and its occupants.

The expansion of RAF bases also had to take account of public concerns over the likely impact on local communities and the environment. This was the context in which the Prime Minister, Ramsay MacDonald, instructed the Royal Fine Arts Commission to become involved in airfield design. A process of consultation with the Air Ministry was initiated with visits by three distinguished architects, Sir Edwin Lutyens, Sir Reginald Blomfield and Giles Gilbert Scott, and a leading authority on planning, Professor S.D. Adshead, to Upper Heyford and Abingdon in November 1931. This resulted in the creation in 1934 of the new post of architectural adviser to the Director of Works and Buildings, and the submission of many of the early building designs to the Royal Fine Arts Commission for their approval. Subsequently, much of this liaison work with the Air Ministry was handled personally by Lutyens (Dobinson 1997: 107–108). The buildings erected during the expansion of the 1930s were, as a consequence, more carefully proportioned than their predecessors, a clear distinction being drawn between neo-Georgian for domestic buildings and more stridently modern styles for technical buildings. From 1938, new buildings and stations, including Middle Wallop and West Malling, made increasing use of concrete and flat roofs in order to speed up the building process and counter the effects of incendiary bombs. Decontamination centres, with their encircling blast walls, appeared on bases from 1937, and were built with the fear of gas attack in mind (Francis 1996: 186–193).

In 1936, Air Chief Marshal Sir Hugh Dowding was appointed Commander-in-Chief of Fighter Command and proceeded to put in place an additional infrastructure that ensured the survival of Fighter Command in 1940. Vital to the new system of command and control were the Chain Home radar stations, the first five of which became operational in 1938 (Fig. 31); Observer Corps posts linked by telephone and teleprinter to the Filter Room at Fighter Command Headquarters at Bentley Priory, Stanmore; the operations rooms, which controlled the Groups into which Dowding had subdivided his command; and finally, within each Group, the operations rooms on the principal sector airfields, which controlled the fighter squadrons.

Also of critical importance to the operation of Dowding's airfields, and especially the sector stations, was their ability to disperse and shelter aircraft from attack, ensure serviceable landing and take-off areas, and control movement. The need for dispersal led to the establishment from 1936 of satellite landing grounds

Fig. 31 Chain Home radar station near Dover: one of the first targets for enemy attack in the Battle of Britain. *Photo:* author

and the adoption in 1938 of the principle that the stationing of aircraft around the perimeter could be an effective means of preventing a knock-out blow (Dobinson 1997: 107–108). The development of radio communication, and the introduction in 1938 of the strip principle – the organisation of the flying field into different zones for take-off, landing and taxiing – brought with them an acceptance that movement on the airfield needed to be controlled from a single centre. Hence the increasingly sophisticated designs for control towers, which evolved from the tower design of 1934 to the Art Deco horizontality of the watch office, with meteorological section, in 1939 (Francis 1996: 118–124).

There was also increasing recognition that the wet clay of the vital sector airfields built around London – Biggin Hill, North Weald, Northolt and Kenley – could pose a serious obstacle to effective air defence. The wet winter of 1936–37 had led Dowding to warn that '[if] our fighter aerodromes were out of action for half a day it might have the most serious consequences' (cited in Dobinson 1997: 155; see also Smith 1989: 11, 19). In March 1939, the Air Ministry eventually agreed to Dowding's proposals for all-weather runways and perimeter tracks to dispersals (Dobinson 1997: 166–168). In the following month it was agreed that fighter stations should have dispersals for three squadrons of 12 aircraft each, after which fighter pens with blast-shelter walls and internal air-raid shelters were erected on key fighter airfields.

Additional to the airfields and radar stations under Dowding's operational command were anti-aircraft and searchlight batteries, acting in unison with barrage balloons. The heavy anti-aircraft gunsites were often substantial constructions including emplacements, living quarters and technical and operational buildings. There were three types of heavy anti-aircraft gunsites: those for static weapons (mainly 4.5 and 3.7 inch); those for 3.7-inch mobile guns; and those accommodating 5.25-inch weapons: At least seven formal designs are known to have been issued for 4.5- and 3.7-inch emplacements down to 1945. Some 980 heavy anti-aircraft gunsites were constructed during the Second World War, mostly positioned close to naval bases, major towns and munition factories. Decoy sites, conceived as an additional means of airfield defence by the Air Ministry in the wake of the Munich crisis, were also constructed to divert the enemy bombers away from their main targets.

It would be easy to forget that, while critical battles for command of the skies were raging overhead during the summer of 1940, a vast effort was being made to construct anti-invasion defences (Dobinson 1996c; Foot 2006). The home defence strategy of General Sir Edmund Ironside, Commander-in-Chief of Home Forces, was based on maintaining a 'coastal crust' of beach defences, combined with static defended lines extending inland over a wide area of the country. Their purpose was to contain the advance of an enemy from the coast or an inland airborne landing by the use of obstacles and troops on the ground, thus allowing time for relief by a mobile reserve. The pivot was the GHQ line, employing most of the existing anti-tank guns in association with anti-tank obstacles, and following, where possible, topographical and man-made features such as rivers, canals and railway embankments. It was designed to shield London and the principal production centres in the Midlands, and was supplemented by a series of command, corps and divisional stop-lines to confine, break up and delay a German advance from the coast. These stop-lines included pillboxes for anti-tank guns or light machine guns, normally combined with roadblocks and weapon positions in the form of trenches and pits. Where there were no natural or man-made obstacles, massive anti-tank ditches were dug. Strongpoints were concentrated at strategic points to create 'anti-tank islands' or 'hedgehogs'.

This policy of fixed lines of obstruction and defence was countermanded in August by Ironside's successor, General Sir Alan Brooke. Greater emphasis was then given to mobility rather than to static defence, which some senior officers regarded as fostering a 'siege mentality'. In October, by which time over 14,000 pillboxes had been constructed along stop-lines, Brooke ordered that the dwindling supply of cement be concentrated on the completion of beach defences. Fortifications came to be concentrated on nodal points, supplemented after 1941 by new anti-tank weapons such as the spigot mortar. The actual threat of invasion was lessened by Hitler's advance into the Soviet Union from June 1941, and in February 1942 Home Forces forbade the further construction of pillboxes. All told, some 20,000 pillboxes had been constructed, together with hundreds of miles of anti-tank ditches and obstacles.

Finally we come to the many crash sites resulting from the battle. It is not possible to give a precise figure for the number of military aircraft lost over Britain, or within its territorial waters, during the Second World War. Some figures do exist for specific areas. Over 1,000 wartime crashes are estimated in Suffolk and 767 in Lincolnshire (McLachlan 1989). The number of aircraft lost at sea is particularly difficult to gauge, but the log of the Skegness lifeboat is a useful source, recording 61 occasions on which the boat was called to aircraft crashes (Finn 1973: 113–114). These crash sites have been the subject of much work over the years by aviation archaeologists, excavating them under license from the Ministry of Defence, and some of the aircraft recovered have been restored. A Hawker Hurricane recovered from Walton on the Naze in 1973 is now displayed at the RAF Museum, Hendon, while a Messerschmitt 109 was taken from the sea off Dymchurch (Kent), in 1976 (Ramsey 1996: 400, 677).

Material Remains and Their Protection

Before we turn to a more detailed discussion of some of the Battle of Britain sites and buildings which remain, it is necessary to review briefly some of the options available for their preservation (see Hunter and Ralston 2007 for a review, but noting again that the heritage protection regime in England and Wales is under review). It has to be decided which elements of the historic environment are to be preserved as found, which can be subject to limited change and which can be exchanged for other benefits. We should remind ourselves that, under the present system, where statutory protection is considered appropriate, the form of protection selected is designed to encourage the type of management that will best ensure the site or structure's long-term future. Scheduling, for example, can be used under the terms of the 1979 *Ancient Monuments and Archaeological Areas Act*. Here the preferred option is the retention of sites as monuments not in everyday use. Listing, by contrast, will be more appropriate where the continuing or new use of built structures is both desirable and feasible. In recommending sites for scheduling or buildings for listing, the role of English Heritage is at present advisory, and each recommendation made to the Department for Culture, Media and Sport must be compelling and demonstrate the site's *national importance* (in the case of scheduling) or the structure's special interest (in the case of listing).

Where statutory protection is not appropriate, the government has issued planning guidance. This includes stipulation that archaeological remains should be recorded in advance of redevelopment or removal. Similar instructions have been drawn up in relation to historic buildings in order to protect the fabric of the landscape. The views of local communities and interest groups also need to be considered. Conservation areas, designated by local authorities, can play a significant role in conserving important sites.

Of all the sites connected with the battle, none has greater resonance in the popular imagination than the sector airfields which bore the brunt of the Luftwaffe onslaught. In the words of Churchill, they were the bases 'on whose organisation and combination the whole fighting power of our Air Force at this moment depended' (1949: 292). The 11 Group, commanded by Air Vice-Marshal Keith Park, occupied the front line of the battle. Its 'nerve centre' at Uxbridge and sector stations at Kenley, Northolt, North Weald, Biggin Hill, Tangmere, Debden and

Hornchurch took some of the most sustained attacks of the battle, especially between 24 August and 6 September when these airfields became some of the Luftwaffe's prime targets.

Biggin Hill (Fig. 32) is commonly regarded as the most significant of the 11 Group's sector stations: more enemy aircraft were destroyed by squadrons based there than at any other station during the war. Following its use during the First World War, work on rebuilding the station in permanent fabric began in 1929, with several buildings bearing the date stones of 1930 and 1931: the Air Estimates for 1933–34 reveal that £190,000 had been allocated for this purpose. The airfield also witnessed pioneering air-to-air and ground-to-air experiments in radio communication and, from 1936, it was used as a laboratory for the Fighter Direction Organisation, which linked radar to aircraft (Hastings 1976: 60). The autumn of 1939 saw the construction of a tarmac runway, measuring 4,800 by 150 ft, and in June 1940, the completion of 12 fighter pens positioned beside the new perimeter track.

The parts of the site which are now missing were destroyed in the Luftwaffe raids of 1940. The raid of 30 August resulted in severe damage to the barracks, WAAF quarters, workshops and stores. The following day, the Sector Operations Room took a direct hit and the hangars were badly damaged. On 6 September, after further raids had rendered much of the base unusable, the last remaining hangar was destroyed on the orders of the base commander (Collier 1966: 190–202; Wallace 1975).

The surviving buildings on the so-called North Camp site are largely representative of the designs associated with Trenchard's expansion scheme. The barracks and station headquarters have all survived remarkably well. Of the buildings which have undergone some alteration, commonly in the form of replacement doors and windows, the most historically important are the station sick quarters, dating from

Fig. 32 Dispersal hut at Biggin Hill, from where Battle of Britain fighter pilots would be 'scrambled'. *Photo:* author

1930, with a decontamination annex added later, and the institute building. Other buildings, such as the motor transport sheds and workshops, also survive.

Across the road that divides the site stand the married quarters, typically planned and designed on Garden City principles and now restored to an excellent standard as private housing. Also to be seen is the officers' mess, a finely detailed neo-Georgian composition now restored as a house and situated next to the telephone exchange centre which played a crucial role during the battle. Although documentation has not yet been traced, it is very probable that the mess was one of the buildings designed in partnership with the Royal Fine Arts Commission during the winter of 1934, the great attention paid to its composition owing much to the fact that Biggin Hill's proximity to London made it the subject of frequent visits by officers from other air forces.

The airfield has retained two fighter pens, which are still in use for light aircraft, the rear wall of a third and a centrally sited Picket Hamilton fort. To the east side near one of the pens there still survives a modest weather-boarded hut, of Air Ministry Sectional B-type design, re-sited and now wedged between two recently built warehouses (Fig. 32). This was the dispersal hut, familiar from wartime photography, as the building in and around which pilots rested before being given the order to 'scramble'.

After Biggin Hill, the sector station to have retained most of its original built fabric is Northolt (west London). The airfield has been subject to considerable post-war development, but many designs of the 1920s have survived from its rebuilding as a fighter station under Trenchard, including the officers' mess of 1923 and the original four barracks blocks. Also to be seen are two hangars, the station workshops and the operations room, and a single pillbox (Norris 1996).

Kenley also retains a few of its buildings. The officers' mess, prominently sited on the west side of the aerodrome and still displaying the scars of the devastating Luftwaffe raid of 18 August 1940, now stands as the most impressive surviving building dating from the rebuilding of the station between 1931 and 1933. Less prominent are the sergeants' mess and the workshops, sold to developers by the Ministry of Defence in 1999 but still standing. Kenley, however, can boast of the most complete surviving fighter airfield associated with the Battle of Britain. A large part of Kenley Common, still managed by the Corporation of London, was converted for use as an aerodrome for the Royal Flying Corps in 1917 and enlarged through an Act of Parliament in 1939. The 800-yard runways and perimeter tracks were completed in 1939, and extended by a further 200 yards in 1943. All 12 fighter pens under construction in April 1940 have survived (Corbell 1996).

Although the airfield at North Weald was remodelled for jet fighters in the 1950s and only one hangar, two watch offices and the officers' mess have survived well from the 1920s, it has the most complete set of fighter pens after Kenley (Nicoll 1996). Debden has retained much of its 1930s character, the best surviving example of which is the operations block. Much of the flying field and defensive perimeter are still intact.

As for the other sector stations, only fragments now remain of Tangmere, where the raid of 16 August 1940 caused great damage and destruction. The deserted and ruinous control tower survives as a lonely icon on the edge of the original flying field, now in agricultural use (Beedle 1996) – see Fig. 10. Very little also remains of Hornchurch, though glimpses of it are still to be seen in *The Lion has Wings,* Michael Powell's propaganda film of 1939. Havering Borough Council has developed what is left of the airfield as a country park, with a fighter pen and some gun posts and pillboxes integrated into a walk around the former perimeter. The former married quarters and officers' mess are now set amongst more recent developments (Sutton 1996).

Not much is left of satellite stations such as Manston and Hawkinge. At West Malling, however, the anti-aircraft/observation tower on the airfield's perimeter has been retained as a landmark at the entrance to a new housing development and there are plans to convert the control tower into a community centre. The former officers' mess currently serves as the offices of Tonbridge and Malling District Council.

Duxford, the most southerly airfield of 12 Group under the command of Air Vice-Marshal Trafford Leigh-Mallory, is one of the best-preserved examples of an interwar military airfield in Britain, with fabric representative of both expansion periods. It was one of only 45 stations retained for the RAF after 1918, first as a Flying Training School and then as a fighter station for 19 Squadron. Duxford, which became famous for its association with the 'big wing' tactics of Douglas Bader, was subsequently chosen as one of the locations for the film *Battle of Britain,* one scene from which involved the destruction of one of its wartime hangars. It was also the subject of a public inquiry in 1976 when Sir Douglas Bader argued for the retention of the entire airfield in opposition to the construction of the M11 motorway across the eastern boundary of the site. It is now the home of one of Europe's leading aviation museums, run by the Imperial War Museum (Raby nd, 1996; Ramsey 1978).

The other aspect of the battle that has usually attracted great public interest is the system of command and control. Pre-eminent among the surviving Chain Home radar stations is Bawdsey (Suffolk), the site of a large Victorian house close to the Deben estuary, where from 1936 Robert Watson-Watt and his team carried out the key experimental work that was to form the basis of Britain's air defence system. A measure of the site's importance was the attention paid to its protection against attack in the summer of 1940, clearly visible today, for example, in the ring of pillboxes on its landward side. Until recently the site was still dominated by the transmitter mast, whose height enabled the transmission of long wavelength signals, but which had become the lone and slightly truncated survivor of the original group of four transmitter masts, their position marked on the ground by large concrete anchor blocks. Alongside the remnants of this mast, the receiver block and the transmitter block – the last, remarkably, retaining its original switchgear – are still to be seen (RCHME 1995).

The Observer Corps have not left any distinctive sites from the Battle of Britain period: their posts were not given permanent structures until the reorganisation of the service in 1942 and, besides the construction of dugouts, huts and shelters on an ad hoc basis in 1940, were chiefly recognisable by the telegraph poles which linked them to the Fighter Command system (Wood 1996). Some excellent examples remain, however, of the operations rooms, which represented a critically important link in the chain of command. Among them is the underground operations room, manned by Dowding and his team in 1940, at Bentley Priory, Stanmore, the late-eighteenth-century mansion which served as the headquarters of Fighter Command. The room was later remodelled during the Cold War period.

The underground operations room at RAF Uxbridge, built in 1938, which became the strategic centre of 11 Group's operations during the summer of 1940, has survived complete with its original air filtration system and power supply (Fig. 33). It was here, on 15 September 1940, that Churchill witnessed Air Vice-Marshal Keith Park deploying the squadrons of 11 Group. The plotting room has now been restored by the Ministry of Defence to match Churchill's detailed description, complete with its 'tote board' and plotting table (Churchill 1949: 293–297).

Debden provides a particularly good example of an operations room at a sector station in 11 Group, comprising a protected roof design of 1937 and surrounding blast walls of reinforced concrete (Francis 1996: 46–48). Duxford, by contrast, incorporates the best surviving example of a 1924 design, which resembles a rectangular, hipped-roof bungalow protected by blast walls made of earth. The operations room at Northolt, though externally altered by new windows and the removal of its blast wall, has retained its original internal plan, with a shuttered opening providing communication between the direction-finding room to its right, where the cross-bearings from the fighter pilots' radio telephones were translated into a map position, and the plotting room.

Fig. 33 Plotting table in the Operations Room at RAF Uxbridge. *Photo:* English Heritage. (NMR BB000654)

Whilst these sites and buildings deserve protection for their place in the history of national defence and their evocation of a heroic stand against all odds, they also deserve consideration for their importance to our understanding of the history of technology and warfare. English Heritage's aim is to broaden awareness of the wider historic significance and educational potential of these sites, in order that ill-considered demolition and alteration should be prevented. It will then be possible to preserve them for the benefit of a wide constituency of interests, from historians and archaeologists to school groups and local communities.

Postscript

Since this chapter was written, many buildings have been designated through listing and airfield defences through scheduling. Kenley has been designated as a conservation area. Comparison with European sites aided the final phase of interpreting the significance of military aviation sites in England (see Lake et al. 2005).

Chapter 11
D-Day Preparatory Sites in England

Between midnight on 6 June (D-Day) and 30 June 1944, over 850,000 men landed on the invasion beachheads of Normandy, together with nearly 150,000 vehicles and 570,000 tons of supplies. Assembled in camps and transit areas over the preceding months, this force was dispatched from a string of sites along Britain's coastline between East Anglia and South Wales (Dobinson 1996b: 2). This short chapter describes those monuments surviving in England which represent the preparations and embarkation for the Normandy invasions of 1944 (see Dobinson et al. 1997 and earlier chapters for a summary of the wider project of which this study forms a part).

Contrary to what has been said previously (e.g. Wills 1994), much of this archaeological record does survive including examples of all types of site constructed or adapted to serve Operation *Neptune* – the assault phase of *Overlord* – which represented the springboard for the Allied invasion of German-occupied Europe. However, there are variations in the quality and extent of survival. Some classes of monument are characterized more by ephemeral remains (camp sites, training facilities); other classes survive in more substantial form. These include: construction sites for the artificial concrete 'mulberry' harbours, and some components of the harbours themselves; repair, maintenance and construction sites for the many vessels involved in the Operation; and the embarkation sites from which troops departed and *matériel* was despatched for the French coast. It is these most obvious and substantial of remains that form the basis of this assessment, though accepting that examples of other monument classes do survive. For bombing decoys, put in place to confuse enemy reconnaissance, assessment has already been completed (Dobinson 2000a: 177ff), while some work has been undertaken on surviving storage and supply depots (e.g. Francis 1997), on sites associated with PLUTO – the 'Pipeline under the Ocean' (Searle 1995), as well as by English Heritage on training areas and airfields.

J. Schofield, *Aftermath: Readings in the Archaeology of Recent Conflict,*
DOI: 10.1007/978-0-387-88521-6_12, © Springer Science+Business Media, LLC 2009

Preparations for Embarkation

Those monument classes representing the three principal aspects – or 'teeth' – of the Operation display some of the most obvious and monumental remains, symbolizing the scale and international significance of the events of June 1944. The three classes can he characterized in the following terms:

Mulberry Harbour Construction Sites

The construction of the two artificial 'mulberry' harbours, built in sections (and different component parts generally at separate sites) and towed across the channel for disembarkation of troops and landing of supplies, was, in Churchill's words, 'a principal part of the great plan', and was decisive in the first days of the invasion. Although one harbour failed, the remaining structure – at Arromanches – was significant in providing the tactical advantage of surprise, and the logistical advantage of not having to land on a defended shore and at the mercy of the weather. Some components of the harbours were clearly surplus to requirements and remained in the UK; some sank on route, or were beached for other reasons. Many sites were involved in this construction process, stretching at least from Southampton, via south coast ports and London, to the northeast.

   Mulberry harbour construction sites were designed variously for the manufacture of Phoenix caissons (partly submerged breakwaters made of cement; the largest was 60,447 tons) and Bombardons (floating steel breakwaters; up to 1,000 tons) which made up the outer harbour, and the pierheads (Spuds), floating piers (Whales) with their steel-spanned roadways, and pontoons (Beetles), some of steel, some of concrete, that supported them (Harris 1994; Hughes 1994). These construction sites were located either in largely unmodified dry docks or slipways, or in excavated basins or on beaches. Much use was made of existing facilities. In Southampton, No. 5 Dry Dock and adjacent wet berths were used to build 12 of the largest Phoenix caissons, while Bombardons were assembled in No. 7 Dry Dock and on adjacent quays and land, the parts coming from all over the country (see Peckham 1994: 13–17 for photos). It is the beach construction sites, however, that retain most evidence for this construction task (e.g. Lepe, Stokes Bay and Hayling Island, all Hampshire), comprising construction platforms, slipways and winch-house foundations (Hughes 1994). One example of an excavated basin is at Clobb Copse, on the Beaulieu river, Hampshire (Cunningham 1994: 181), used for the construction of a Bombardon breakwater and 50 of the 470 Beetles constructed for Overlord. At Marchwood military port near Southampton, Beetle and Whale units were assembled on rail tracks and moved using a traverser or turntable. This traverser survives within the modern port. Some of the components built for the harbours also survive, mostly at sea, having sunk while on tow (e.g. Phoenix caissons in Portland and Langstone Harbours, and near Southend) but occasionally on land, as with the line of Beetles at Dibden Bay (Hampshire). Sunken mulberry debris has been noted by recreational divers at various locations off the south coast (McDonald 1989; 1994; Pritchard and McDonald 1991). Of course, parts of the mulberry harbour at Arromanches survive in situ.

Maintenance and Repair Areas

The maintenance and repair areas, and harbours used for landing craft and landing ship construction, were essential to developing and retaining a fleet capable of delivering Churchill's 'great plan'. With so many vessels involved (landing craft and landing ships principally, but there were 46 different types of vessel in all), maintenance was a significant task. Contemporary descriptions talk of unprecedented levels of maritime activity, with every port, harbour and boatyard being involved, in addition to beaches, on specially built slipways and gridirons, and in the streets of coastal towns and villages.

The purpose-built gridirons (also known as 'scrubbing grids') were used for maintenance, and took the form of a series of parallel concrete rails running down a slight gradient into the water, allowing a boat to be floated on at high tide, and repaired at low tide; some were supplied with a winch mechanism for pulling vessels onto the grid, and timber and steel mooring points ('dolphins') for securing them when afloat. Recorded examples are confined to the Rivers Dart (Devon), Tamar and Fal (Cornwall) and Portsmouth Harbour. At Lower Noss on the River Dart, one example survives with two sets of gridirons, each with concrete mooring posts, and steps leading up the river cliff behind to a levelled area with hutting and metal racks, presumably representing workshops. At another site, Mylor (Cornwall), archaeological evaluation has shown that part of this gridiron at least sits on a concrete raft, while the testimonies of those involved suggest that the shuttered concrete rails each supported a securely fixed timber rail on to which the boats were hauled.

Repair areas in the form of slipways, with a metal rail, winch mechanisms and dolphins, are known to have been used for landing ship repairs. Examples survive at Mill Bay (near Salcombe) and Waddeton (both Devon). The Mill Bay example is particularly well preserved (Fig. 34), and has the benefit of appearing in contemporary photographs with a landing ship in situ (Murch et al. 1994: 9). However, much of the repair and maintenance activity was conducted on an ad hoc arrangement and leaves little trace: for example, landing craft (assault) – LCAs – were small vessels constructed and repaired mainly in back streets and improvised hards at the water's edge.

Embarkation Sites

Embarkation sites had to be well-designed and well-built if embarkation was to be a rapid and efficient exercise. Geographically the sites had to have access to hinterlands in which large numbers of troops and supplies could be concealed from enemy

Fig. 34   Slipway at Mill Bay, near Salcombe (Devon), seen at low tide. *Photo:* author

reconnaissance, yet which had the road and rail networks to allow ease of movement at the time of departure. This part of the Operation was planned well in advance, with most embarkation hards built in the period October 1942 to spring 1943. In all, 68 embarkation sites are documented in public records (Dobinson 1996b), representing those built specifically to serve general cross-Channel operations from 1942 onwards, and the extension to that group built to serve Operation *Neptune*. The list is complete in both these respects though, as photographs show, embarkation also took place at other sites not built for the purpose.

   Embarkation sites were either modified docks, quays or harbours (such as Southampton Docks) or were constructed specifically for the purpose. Two main types of loading facility were used: LCT hards for 'landing craft, troops' and LST hards for 'landing ship, tanks'. Although LST hards were the most numerous, the two types were broadly similar. Each had: a concrete apron (solid concrete above high water, and flexible concrete matting below), and a series of dolphins; hutting for offices, workshops and stores; fuelling facilities; electric lighting and roads and transit areas (see Dobinson 1996b: 14–18 for details). Survival tends to be confined to those hards built specifically for the purpose (those in existing docks having been redeveloped in the post-war period): Torquay (Torbay), Brixham (Devon), Turnaware, Polgerran Wood and Polgwidden (Cornwall), Lepe (Hampshire), Stone Point and Stansgate (Essex), and the hards at Upnor (Kent) are among the best.

   At the LST hard at Turnaware, the concrete apron survives along with two mooring posts, and the fragmentary remains of dolphins and hutted accommodation.

The approach road, which divides, presumably to allow a one-way system to operate, also displays what may be contemporary vehicle tracks (Fig. 35). Of the LCT sites, Torquay is the most substantial, representing an outstanding and monumental example of D-Day architecture (Fig. 36). It was from here that the American 77th Infantry Division embarked, destined for Utah Beach. Importantly, these two embarkation slips survive dominating the modern harbour, and serving as a focus for the commemoration services of Normandy veterans. These structures were listed at Grade 11* (listing is a form of heritage protection) on the 56th anniversary of embarkation, 6 June 2000.

Finally, mention should be made of PLUTO, the pipeline under the ocean, and some of the camps established and occupied in the months prior to embarkation. PLUTO (and SOLO – the pipeline under the Solent, a crucial link in the network) was established to provide fuel for the invasion force, and took the form of a complex and extensive system of pipelines and terminals, with pumping stations at Dungeness (Kent) and Sandown and Shanklin on the Isle of Wight (see Searle 1995 for details). Although the pipelines were cleared from the seabed after the war, short sections do survive, particularly off the Isle of Wight. A pipeline valve survives at the Hamble oil refinery in Hampshire, while at the SOLO terminal at Thorness Bay on the Isle of Wight, shore-end pipe connections are exposed at low tide (Searle 1995: 84).

In most areas, little is likely to survive of the many camps occupied prior to embarkation, though preservation is likely in places like the New Forest. What does

Fig. 35 A moment in time: detail of the LST hard at Turnaware (Cornwall) showing what might be contemporary vehicle tracks on the road surface. *Photo:* author

Fig. 36   The slipways at Torquay (Torbay). *Photo:* English Heritage (AA012294)

survive, however, are the sources which reveal their locations and the effect of this encampment on the contemporary landscape. Aerial photographs by the United States Air Force (March 1944) for the Truro area show the hundreds of bell tents occupied by US troops concentrated along arterial routes (Nick Johnson, personal communication). Contemporary maps and plans, and ground photographs, show the overall layout and the character of these sites (see various photos in Doughty 1994). Along the minor roads linking Lepe with the New Forest, extended lay-bys date from this period while contemporary photos show vehicles using them, parked up under the trees. Finally, many of the fuel dumps, hutted camps and hospitals do survive, though often now as developed sites: industrial estates, modern hospitals and garages.

The Future

In terms of the three principal aspects of the Operation, therefore, significant archaeological remains do survive. While the view that static defences will survive better than structures and sites representing a mobile offensive may be valid in general terms (Wills 1994), sufficient remains of the preparations for D-Day in southern England to give an impression of the scale of the Operation, and the variety of the specific tasks involved. For these reasons, D-Day sites have an international significance, alongside the battlefields of Normandy, in representing the physical manifestation of arguably the most significant military event in European – perhaps even world – history. Therefore, sites in England are considered significant where: surviving remains constitute a particularly rare example of a structure or component; sites survive in close proximity displaying the intensity and scale of the Operation, as witnessed at a local level; the range of components present combines with a degree of visual integrity to provide for a full interpretation of the site and how it functioned; the site displays a high quality of surviving remains; the size of the site is indicative of the scale of the Operation; and this can be appreciated in a physical setting which has changed little over the last 50–60 years.

The events in Torquay on 6 June 2000 demonstrate the validity of this approach, and the need to retain such sites for reasons of education, commemoration and remembrance. As one veteran said after the commemoration service, and after hearing news of the Torquay slipways' protection: 'The listing is terrific news – [they] are just about the only thing left to tell future generations about what went on at the time.'

Chapter 12
Le Carré Landscapes: The Cold War

This chapter presents a critique on Cold War heritage, describing why and for whom it has relevance and why archaeologists in particular appear to engage the subject with such enthusiasm. It also addresses the diversity of this historic legacy and the range of values that are ascribed to it. It would be wrong to describe this legacy as straightforward. This is an archaeology of discord which presents difficulties and challenges at every turn, but perhaps that is the attraction? This chapter will attempt to convey something of the climate that accompanies this cultural material: the heat that accompanies Cold War material culture and its conservation.

Background

A comment sometimes made about archaeologists' developing interest in twentieth-century heritage is that archaeology is about distant times, commonly time before history, time during which there are no historic records to draw upon and no testimonial evidence. But archaeology is more than this. It is an approach to studying material culture, a perspective on the past, and as an approach this is just as valid for recent times as it is for prehistory. Indeed in some ways the Cold War sometimes *is* prehistory, as records and documents can be closed or 'classified' (at least they are in the UK) and participants remain bound (or think they are) by the *Official Secrets Act*. As an example, documents have not yet been systematically studied for Cold War civil defence. The fact that so many civilians would have died while a select few local and national government officials survived, presumably to satisfy themselves of some semblance of order post-apocalypse, is something only made obvious by close scrutiny of the material record. Only by visiting and examining bunkers from an archaeological perspective are these implications revealed. An archaeological study of local and national government defence provision will focus on the numbers of rooms within bunkers, the signage and labels on doors, revealing who would occupy each office, the types of equipment and furniture contained within and the proportions of sleeping to office accommodation and facilities. For private shelters, we may wish to know how many were built and for whom. To date very few examples have been reported, suggesting that few were built. With all of this information to hand, it is possible to establish how the system was intended to operate, how many people could be accommodated and for how long, and who, and how many, would have survived? The documentation that might provide answers seemingly isn't available for us at the present time.

It is likely that everyone currently involved in conservation in a professional capacity will have experienced the Cold War and know its history to some extent, but a short historical review will provide context for what follows. The Cold War followed the close of World War II and was characterised by the conflicting ideologies of East and West which led to an escalation in the arms race and increased militarisation around the world. In the years 1946–89 this was a world in which the 3-min warning was a constant threat and the shadow of the mushroom cloud an enduring image. This was a time in which peaceful protest challenged authority, with the Aldermaston marches followed by the permanent residence of women at Greenham Common (Taylor and Pritchard 1980). This was also a time in which fear, suspicion and the emergence of a big brother watching our every move, either from within or by enemy reconnaissance, was increasingly reflected in popular culture. The traces of Le Carré landscapes can still be seen and felt in Berlin ten years after the Wall came down. Cold War films still have cult status and some continue to be produced. Cultural tourism reflects immense popular interest in the period and popular music of the time took mutually assured destruction as a recurring theme.

Set against that backdrop, this chapter will seek to unravel or 'excavate' legacies of the Cold War in four related parts. These parts will:

1. Briefly outline the assessment of Cold War sites in England, its scope and methodology and the challenges that face those that manage Cold War sites
2. Describe how the division between East and West is represented in the material record and just how far the influence of the Cold War extended
3. Look at the inspirational qualities that Cold War sites have attained and the opportunities these qualities provide
4. Review some of the values that can be attributed to Cold War places

Part 1

Following initial emphasis on World War II (see Chaps. 1–3), English Heritage has moved on to study the Cold War. Unlike the work on World War II, primary sources were not so readily available for this period though those that were available, for Bloodhound and Thor missile sites, the Royal Observers Corps and Rotor (a post-war development of radar), were examined. More traditional archaeological methods were adopted for this project instead as the means to achieve an assessment of surviving sites, building on an earlier recording project undertaken by Roger Thomas and Wayne Cocroft of the (then) Royal Commission on the Historical Monuments of England. Their brief was to locate and record exemplars of the major classes of monument from this period (Cocroft and Thomas 2003). Following completion of this recording project and following the RCHME's merger with English Heritage, Wayne Cocroft undertook a subsequent assessment of Cold War monuments in England (Cocroft 2001; James 2002) based on the earlier recording project and a separate study of the explosives industry (Cocroft 2000).

This assessment involved, first, establishing a chronological context within which sites could be understood. Research highlighted two key phases of the Cold War, the first from 1945 to 1962 which from 1950 onwards was characterised by a massive rearmament and defence building programme; the second was from 1980 to 1989, being a period in which sites were increasingly hardened and built up. The period in between was one of sustained deterrence which during the late 1970s involved the implementation of the programme to harden frontline bases. Within this context, the typology established for the earlier project could be refined and developed and, so far as was possible, sites could be located and documented; what remained of these sites could be assessed and future management recommendations made. In all, eight separate categories of monument were identified, containing 31 distinct monument classes, some of which subdivide further into individual (and still distinctive) types. To give an example: the category Air Defence divides into five classes – Radar, Royal Observer Corps, Anti-Aircraft Artillery, Surface to Air Missiles and Military Airfields. Each of these can be further subdivided by type.

Unlike sites of World War II, it is difficult to link Cold War sites in England to specific historic events such as the Battle of Britain. Instead, criteria can more appropriately reflect political alliances and the changing military strategy that set this era apart. For instance, one of the results of this developing technology was very rapid infrastructure obsolescence. Central to all Cold War planning was the possession by the Superpowers of vast arsenals of nuclear weapons. The development of nuclear technology for peaceful and warlike purposes was also one of the most significant post-war scientific and industrial advances. The development and deployment of nuclear weapons is one of the key characteristics of the Cold War. So structures that were designed to develop, test, assemble and handle nuclear weapons and to operate within a post-nuclear attack environment are deemed to be the most distinctive and characteristic monuments of this period. This category includes for example features and structures associated with the British V-bomber force that was central to British nuclear deterrent policy from the late 1950s–60s and with NATO cruise missiles during the 1980s. Technological significance has therefore been a major factor in this assessment process. As well as being Cold War structures, many sites including those associated with the intermediate range missile Blue Streak are important monuments to post-war British achievements in science and technology.

Beyond Greenham Common (discussed in Chap. 7) are the significant Cold War sites at Alconbury and Upper Heyford (Schofield 2002b), where discussions are continuing to seek a mutually acceptable solution in which these vast areas of immense symbolic and cultural value can be developed and used without unduly compromising their historic integrity. In most cases this will involve a combination of measures: the protection by designation of some key components, perhaps including the Victor Alert Squadron area at Upper Heyford (Fig. 37) and the Cold

War era control room, while retaining the site's overall plan form (for instance ensuring that housing development respects the original layout if not architectural styles from the USAF era) and perhaps street names too; and at the widest scale ensuring key sight-lines are retained, for example by aligning the main arterial route of any new settlement along the line of the main runway, illustrating the scale of its imposition on the pre-war landscape. Here also some of the characteristic hardened aircraft shelters which housed F111 bombers have been converted, with minimal alteration, for office and storage use. In some cases doors have been propped open and a new glass front and a floor inserted. All of these changes are reversible and barely alter the outward appearance of the shelters, respecting their functional form.

Fig. 37 The Victor Alert squadron area, at Upper Heyford (Oxfordshire). *Photo:* English Heritage (NMR 18537/18)

The Cold War presents many challenges therefore, but these are challenges that developers and local communities appear prepared to accept and resolve. In Lincolnshire, the American presence had a profound influence on local culture and identity in World War II and later. This is reflected in the community's enthusiasm for this heritage and a desire to see it retained to some degree at least. The local authority in Lincolnshire has produced an airfields trail leaflet and was at one stage considering a bomber landscape project which would examine the physical and social impact of military aviation on the county.

Part 2

The Cold War encapsulates many different political, social, cultural and economic agenda. The East/West divide was key to this and is now critical for interpreting and deconstructing Cold War legacies. A microcosm of this division existed in Berlin, where traces of the boundary still remain, though those traces are at best subtle and typically intangible – representing a landscape of memory. Little of the Wall itself remains and the distinctive architecture of East and West is now largely blurred by post-1989 development. But both in Berlin and across Europe more widely material cultures of East and West are still there to be studied, analysed and examined. For instance: statuary from the Soviet era in Berlin (Ladd 2002); the archaeological traces and social legacies of the Wall (Feversham and Schmidt 1999); histories of mourning in East and West (Ludtke 1997); the impact of music at the end of the Cold War (Pekacz 1994); and domestic consumption (Buchli 2007). A recent survey of land use change in Russia during the 1990s has shown how the agricultural system has drawn closer to its western counterparts (Ioffe and Nefedova 2001), a point reflected in Otto Braasch's aerial photographs of areas either side of the frontier in the 1980s and 90s. Related to this are changes in Poland where new farming strategies provide opportunities to implement improved policies and systems aimed at the protection of the cultural landscape (Szpanowski 2002). Deconstructing these changes and exploring legacies of the Cold War in this way is fruitful ground for various research initiatives and for collaborative projects.

The influence of the Cold War extended well beyond the front line. Its influence was not always obvious though, sometimes existing only in the undercurrents of social and political life and in popular culture (Fig. 38). As Bill Johnson (2002) has demonstrated, in Las Vegas, which in the 1950s welcomed the nuclear testing programme to the area with great enthusiasm, the cultural response was varied and often radical and reactionary. Houses were designed with large plate glass windows for instance, openly challenging the advice of the Federal Civil Defense Administration which championed more robust construction methods such as those put through their paces at the nearby Nevada Test Site. This influence extends into the post-Cold War world. In Las Vegas, 24 of the world's 25 largest hotels are on the Strip. Those built prior to the Test Ban Moratorium in 1992, which brought an end to nuclear testing at the Test Site some 65 miles away, were substantially built with deep and robust

Fig. 38   Popular culture and the Cold War, Las Vegas. *Photo:* author

foundations designed to survive the earth tremors that accompanied every under-ground test. All large hotels were effectively earthquake-proof. Since 1992 however risk management seems to have become less significant in building design and con-struction. The c. 400-m-high Stratosphere Tower with its rollercoaster at the summit, making it a popular tourist attraction as well as another vast hotel complex, is rumoured to have foundations only 5 m deep – in alluvium. Corners were cut at the design stage and those involved in the Tower's construction are alleged to have said they would never go up it. Local residents have identified an impact zone around the base and parents threaten naughty children with it: if you're really bad we'll send you up the Stratosphere Tower! But it is no joke of course. There is the possibility that nuclear testing will resume at the Test Site (George W. Bush favours this according to some press reports). If that happens, there will be significant impact on the city of Las Vegas which presumably would not welcome a return with the same enthusiasm that it welcomed testing in the first place.

Ideological distinctions exist between those who favoured nuclear armament and those who did not There were many oppositions and disagreements in the Cold War, between political parties, between key individuals within those parties, between con-stituencies (the peace women, the military and local residents at Greenham Common and in addition the traditional owners at military test sites in the USA and Australia) and within each constituency. There were many factions within what was collectively known as the peace movement for example. So it is a heritage that binds many inter-ests and constituencies together and is socially inclusive as a result.

The Cold War therefore was everywhere and affected everyone, whether they really felt its heat or not. That legacy continues in the way we are, what we think about the world, the places where we live, the language we use and so on.

Part 3

Continuing this theme of inclusivity, the Cold War engages many separate disciplines and interest groups, many constituencies in the academic sense: archaeologists, sociologists, archivists, geographers and historians (political, historical and economic), architects and artists for instance. Carefully constituted enquiries into the period can embrace all of these interests in engaging, exciting and challenging studies of the past and of contemporary culture, exploring how we got into the situation that we did, why we were the way we were and are the way we are now, and what Cold War legacies offer the future, for example in terms of cultural benefits. A recent development has been for artists, writers and musicians to use the Cold War as the inspiration for their own work. The Turner prize-nominated Wilson twins were inspired by the missile shelters at Greenham, for instance (Corin 1999). The Trust who manage much of this site employed an artist in residence to record the transition of the former airbase to common land (Kippin 2001) and a book of poetry inspired by Greenham has been published (Symmons Roberts 2001). An abandoned Soviet Base in East Germany has been the subject of a film *Cood bay Forst Zinna* by the artist Angus Boulton (2001, 2007), who has also recorded these places using still photography. Keith Watson uses abandoned bases in the UK as the inspiration for his photography (2004, 2007), and so on.

This then is the third part of this excavation, the inspirational qualities of what for many are dull, unprepossessing, dangerous and menacing structures and sites, qualities that seem to encourage the many diverse and imaginative uses to which these places are now increasingly being put. In 2001, Yannis Kyriakides, a Cypriot composer now living in the Netherlands produced *a conSPIracy cantata*, *SPI* for short (www.unsounds.com) (Kyriakides 2001, 2007). *SPI* is an electronic cantata that juxtaposes two forms of cryptic message communication, the clandestine world of spy number transmissions on the short-wave radio and the enigmatic uttering of the ancient oracle of Delphi. The spy number transmissions sprang up on short-wave radio in the 1960s at the height of the Cold War. They were used to transmit coded text messages in numbers, phonetic letters, morse or noise and were operated by the world's intelligence agencies to relay messages to their agents in the field in anonymous and undetectable form. So, as well as viewing some of the visual images contained in the references cited above and reading the poetry, the third part of this excavation should also (perhaps simultaneously) involve hearing the piece, perhaps Track 2, which contains sung fragments from short-wave radio transmissions on a station nicknamed 'the Czech Lady'. It also includes part of a list of 100 basic words that formed part of the encoded secret messages.

SPI is a soundtrack to archaeological work on the Cold War, influenced by its political machinations and the clandestine world that was created and reflecting the nature of what life, for many, had become. Alongside the protest songs of Bob Dylan, Joni Mitchell and others in the West, the thrash and heavy metal in East and West that yelled of mutually assured destruction, the punks that recorded social and economic disintegration so effectively and the techno music of Berlin's clubs (see below), this is truly the sound of the era, hard to comprehend, difficult to disentangle, but enthralling and captivating all the same. As part of the 2002 Aldeburgh Festival, *SPI* was performed in the so-called Star Wars building (in reality a debrief centre) at RAF Bentwaters, a Cold War USAF base in Suffolk. The sounds of the Cold War returning to and reflecting one of the places that created them in the first place.

Part 4

So what are these Cold War sites beyond the obvious monumental manifestations of this crucial and far-reaching era in recent cultural history? They can be a number of things and in some cases several of the following.

They can be *enspirited places* (after Read 1996). More so for wartime sites, but increasingly for Cold War bases, veterans return to relive what for many was the focus of their service careers: delivery of what the politicians had decreed. The number of veterans' Websites and the sense of belonging and nostalgia they convey reveals, often quite explicitly, feelings of attachment held by retired servicemen. The Nevada Test Site is enspirited for other reasons, by the Western Shoshone, excluded from traditional lands by the Defense Department. On Bikini Atoll it is the Bikini Islanders, also forcibly removed to make way for nuclear tests but now returned to a dangerous and contaminated place which they are effectively exploiting for tourism at great risk to their own health and well-being (Delgado 1996). At Greenham, the Common has now been returned to the people of Newbury who can access and use it once again. Cold War sites are lost places for many. Places where lives were lived, children born and parents and loved ones died. They are settlements, homes like any other, and are enspirited for being so. The wall art that characterises so many USAF sites (the street-gangsta style distinguishing it from British wall art) is a further legacy of that particular connection with place (Cocroft et al. 2006).

They are also often *encoded places*, places that can only be understood by decoding the material cultures that remains, with archives where they exist and sometimes testimonial evidence. Decoding can be a simple process such as interpreting the basic plan form of an anti-aircraft gun site. Examining in functional terms the relationship between access roads, stores, mess and domestic accommodation and the emplacements is comparatively straightforward and highlights the value of retaining that range of components for purposes of presentation and interpretation. Or it can be complicated. The Barnham Nuclear Bomb Stores in Suffolk is an example (Fig. 39). Here the key to understanding the site is an appreciation of the processes and rituals by which nuclear weapons were assembled, managed and moved about, the flow-lines in industrial terms, and the rules determining who could go where, by what means and what they could do when they got there. The speed of the process too is a key aspect of understanding much about the Cold War according to the theorist and philosopher Paul Virilio (Virilio and Lotringer 1997). The significance of the impact of speed on modern life is exemplified by the Cold War and the threatened 3-min warning, in which we would all have had 3 min to do whatever it was we were going to do, or were ordered to do: firing a missile,

Fig. 39 Site discipline at Barnham bomb stores (Suffolk). *Photo:* author

reaching the bunker, saying goodbye. More complicated still are intelligence and research and development sites where experimental work, research, production lines and information flow all require analysis and comprehension. Bletchley Park near Milton Keynes, and the GCHQ buildings at Cheltenham are extreme examples of this, as are Orford Ness on the Suffolk Coast and Foulness in Essex.

Many of these sites are *contested places*. They mean different things to different people, often people who were and may still be in direct opposition to one another. Greenham Common is an example here, one at which opposition can be seen in the physical remains, the fence for example, the caravan, but where further meaning comes from testimonial evidence such as can be heard on the Imperial War Museum's Website (www.iwm.org.uk/upload/package/22/greenham/index.htm). Related to this are sites where different perceptions exist but are not necessarily in contest. These can be sites which represent discordant but often positive values, such as Berlin, a cultural hub where people from east and west (and formerly East and West) now dance to techno tunes in buildings previously occupied by the Stasi. But with the techno music that created the need for these clubs, and which maintains them, likely to wane, what is their future? It is odd to think of these key historic buildings, witnesses to momentous decisions and awful events in recent history, being subject to the transience of youth culture. These adapted structures symbolise the unifying process, yet they would not support any musical traditions other than those which have thrived in them over the past decade. These are no discos – it is their post-industrial hard-edged characteristics that make them what they are. It is the music that now keeps them alive.

The continued presence of sections of the Wall ensures that this era is not forgotten, but more importantly it enables and ensures a debate which otherwise might not occur (though at the Topography of Terror exhibition the Wall runs alongside a site of Nazi atrocities, presenting a confusing message, certainly for those too young to remember both eras). In other words, as with the battlefields of World War I, it is ultimately about not forgetting.

Many of the places mentioned here are or will become *protected places*, either through ownership and the management regimes created for them, as at Orford Ness, whose managers the National Trust have pursued a policy of management through neglect, or as designated sites (currently scheduled monuments or listed buildings) or through management agreements. Whether protection now for the future is correct only time will tell. But we, the heritage community, would be failing in our duty if we did not seek to retain some sites now in order that future generations could also make that decision. By acting promptly and proactively at the end of the Cold War, we have also ensured that a representative selection of sites can be retained. For other periods of history (and for World War II to some extent), we could preserve a sample only on the basis of what had come through the first 60 years, a sample of a sample, and one determined by post-war development and agricultural policies and practice and (in eastern counties at least) rapid and devastating rates of coastal erosion.

Some are *inspirational places*. Writers, artists and musicians have used the monumentality of the Cold War to create their own representations, interpretations and deconstructions of the past. Greenham Common is notable for this reason too.

Some are best known as *contaminated places* – where historical use is considered to have blighted future opportunity.

Finally, it is worth elaborating on two broader themes that hold relevance here. First is the distinction between Cold War-*affected* landscapes (the airfields and training areas: places where the military and its execution of the Cold War geopolitical and military agenda have directly impacted upon the landscape) and those which are Cold War *related* (places like East Anglia where American servicemen had a sometimes subtle and on occasion profound influence on the local scene). Line dancing and Country and Western music in the UK are rumoured to have emanated from the concentration of US bases in East Anglia before becoming fashionable nationwide; American cars still sometimes clog the East Anglian lanes. Some of the Cold War-affected landscapes can be described as being in the front line. Although as we have seen the Cold War was extraordinary in being everywhere and nowhere – a placeless war – there were sites which were undoubtedly 'front line', whether as targets (Cheltenham for example, because of GCHQ's presence) or places from which a pre-emptive strike would be launched (Greenham Common, Upper Heyford or Molesworth, targets as well of course). Mechanisms are in place for managing Cold War-affected landscapes. But options for those that are Cold War related are more limited. Understanding the character and the extent of these areas may be a first stage.

The second theme is the industrial process and the military–industrial complex that we saw emerging in Chap. 9. As was mentioned earlier, there are processes at work within the technological and cultural spheres, processes that often hold the key to understanding and interpreting the sites and buildings themselves, an essential prerequisite to informed conservation. This can be at the local scale of how weapons were manufactured, maintained, assembled and prepared for use. For example, what rituals can be identified through the archaeological record? Or it can be at a wider geographical scale: Blue Streak for example. This was an intermediate-range ballistic missile designed and manufactured in factory buildings at Stevenage, driven (slowly) to Spadeadam in Cumbria, tested on a static test stand there, then transferred by sea and land to Woomera in Australia for live firing. Rockets used in the development of its nuclear warhead were tested separately off the Isle of Wight before they too were shipped to Woomera. All of these facilities survive. Woomera is now the site of a controversial refugee camp; the Isle of Wight site is managed as a monument by the National Trust and Spadeadam is owned by the Ministry of Defence. Cold War heritage is world heritage, without question.

Conclusion

To summarise the results of this excavation, Cold War material culture has huge cultural significance yet historical interpretation is sometimes difficult given the lack of available records and testimonies. However, the complexity of these sites and the meanings and memories they convey provide unique opportunities for an archaeology of 'us', a (re)presentation of this crucial period of history through a methodology that embraces many interest groups and media and a diverse material record: art and archaeology; buildings and monuments; war and peace, either side of a tattered fence; massive concrete bunkers designed to withstand all but a direct hit; the 'Magic Mountain' at Alconbury – Britain's biggest bunker and a caravan that was home to those who opposed the deployment of arms. These together exemplify the extraordinary archaeology of the Cold War.

Section 4
Further Directions

The title of this book – 'Aftermath' – hints at a move away from traditional approaches to military archaeology, the emphasis shifting from a focus on sites and artefacts, the physical remains per se, to a broader archaeology of militarism and its cultural and social significance as heritage: why places and their traces matter and to whom. The three previous sections have all addressed these concerns from slightly different perspectives: establishing management and intellectual frameworks within which more detailed studies can be established (Sect. 1); the very particular connections that can occur between memory and place (Sect. 2); landscapes of events (Sect. 3); and now further directions. The book is about people and places in other words, not just the physical remains that militarism and conflict have left across much of the globe in the twentieth-century.

We can now turn to explore two new directions which, like the previous sections, are not intended as definitive commentary on the future of conflict archaeology, but as a distinct perspective on this particular and troublesome category of material culture. In this section, I include just two chapters, which represent very particular directions that my own work has taken most recently, each from a very different starting point.

Chapter 13 has its origins in a Council of Europe project: 'Responses to Violence in Everyday Life in a Democratic Society', published as a collection of essays in 2004 (Dolff-Bonekämper 2004). The book's Foreword noted how,

> Heritage reflects the periods of openness, peace and prosperity in our continent's past – but it also reflects the periods of tension. If we want to form a clearer picture of the history of European society and the origins of some of the conflicts which have divided it, then we need to consider the various ways in which cultural heritage has been interpreted and the disagreements which it has engendered. … This collection suggests linking the heritage theme with that of frontiers - natural frontiers or frontiers of the mind. Frontiers are critical. One is either on 'this side' or 'that side'. Frontiers are disturbing. They are places of confrontation, expansion or negation. They mark off identities and groups. But they also hold a special fascination, as dividing lines which invite us to strike out in new directions, forge new contacts, and transcend the old and familiar (anon 2004: 9).

'Dividing Lines, Connecting Lines' contained two chapters of mine, only one of which is included here. The other chapter, jointly written with Marieke Kuipers on 'Lines of Tension' (2004) examined defensive lines of the twentieth century: the defensive curtain around Verdun, the Atlantic Wall, the Iron Curtain, and the fence around Greenham Common Airbase. We asked the questions: what is left of these defences, what can their traces tell us, what do they evoke? We also addressed their potential heritage value for Europe.

The chapter reproduced here presented the opposite extreme to the borders that exist between states and power blocs and which left their mark on huge tracts of territory, looking instead at new urban frontiers in London. As Dolff-Bonekämper described it,

> When people who share a nationality, an ethnic group, a religion or sexual preference congregate in a given area, the result is a social and cultural homogeneity, which is reflected in signs, décor and recognisable patterns of behaviour. Recognisable, above all, to those who share the codes, and know exactly where their territory starts and ends. Sometimes obvious, sometimes invisible, these frontiers are social realities in Europe's major cities, and are transforming their cultural topography, particularly in underprivileged areas, where ethno-cultural groups cluster in 'ethnoscapes' – sometimes coexisting, but usually in conflict. (2004: 13)

She goes on,

> Borders are places where we meet others, but they also delimit our 'home'. What is 'home'? Who defined it, when and for what reason? Whose 'home' is it and is it more a home to some than to others? 'Home', as the focus of personal and collective identity, can expand or contract, and exists on many levels – from private to public, local to national. (2004: 13).

The idea of 'dividing lines which invite us to strike out in new directions, forge new contacts, and transcend the old and familiar' also transcends the final chapter. But here the familiarity is something very different. As it states in the text, this chapter originates in a realisation that much of the work archaeologists have been doing in the broad field of conflict archaeology was being replicated by the endeavours of some artists. Archaeologists were recording buildings abandoned by military units and personnel following the end of the Cold War; and artists were doing the same thing, albeit for different reasons and using different media. Indeed artists' and archaeologists' work can often overlap, whether in prehistoric or modern settings. Chapter 14 examines the close connections between artistic and archaeological endeavour using places of conflict to illustrate the nature of this close correlation of interests and motivations. It should not surprise us that this is so, and perhaps it is fitting to have, as the final chapter, a contribution which – more than any other – challenges the disciplinary boundaries that have traditionally existed between diverse research practices. The academe appears increasingly post-disciplinary, and a world where sometimes surprising collaborations are now encouraged by the funding councils. Chapter 14 stands as a small contribution to that debate.

These two chapters are each included by kind permission of the original publishers. The details of the original publications are: Schofield, J. (2004). New urban frontiers and the will to belong. In Dolff-Bonekämper, G. (ed), *Dividing lines – connecting lines – Europe's cross-border heritage*, pp. 69–92. Strasbourg: Council of Europe. This chapter was originally accompanied by a photo-essay by Kristin Posehn. The photo-essay is not reproduced here. Schofield, J. (2006). Constructing place: when artists and archaeologists meet. ebook published by Proboscis: http://diffusion.org.uk/?tag=art.

Chapter 13
New Urban Frontiers and the Will to Belong

Here conflict archaeology is taken in a different direction, emphasising the wider social context beyond militarism and combat. This chapter describes a particular type of tension amongst urban communities, addressing the degree to which borders are constructed and used in the urban environment. It describes how people distinguish the places where they live from the 'other' that lies beyond, and how that separateness manifests itself both to the community and to those from outside. After a background that covers theoretical frameworks and relevant principles of heritage management, three case studies describe three distinct communities in London, with a view to assessing: the methods by which information might be gathered concerning these new urban frontiers; the tangible and intangible heritage that communities call their own (intangible in the sense of heritage without expression through material culture); and the difficulties that can arise where physical boundaries are imposed as a replacement for the hidden or invisible boundaries that existed before.

Constructing Urban Space

Segregated residential patterns within urban space will generally reflect the operation of twin processes of choice and constraint: in some cases, a community may choose to live segregated from other groups; in other cases, processes of prejudice and discrimination may be at work (Moon and Atkinson 1997: 265). The principles and protocols that govern this division of urban space and give it physical expression are well documented, in particular in the fields of human geography and urbanism (Newman and Paasi 1998). Readers will be familiar with the manner in which neighbours set themselves apart with the generally mundane and predictable, sometimes imaginative, and occasionally provocative use of hard (brick, concrete, wood) and soft (hedges and beds) boundaries. Also well documented are the hard boundaries that divide communities within cities, boundaries that may have their origin in political or religious divisions, sometimes deep rooted, sometimes recently formed. The Berlin Wall, for example, is well known, surrounding West Berlin during the Cold War as part of a wider Iron Curtain. As Feversham and Schmidt have said (1999: 10), the Wall stood both as a symbol of the Cold War and a tangible marker of the geopolitical

J. Schofield, *Aftermath: Readings in the Archaeology of Recent Conflict*, DOI: 10.1007/978-0-387-88521-6_14, © Springer Science+Business Media, LLC 2009

division of Europe. Managing the legacy of the Wall as cultural heritage (Dolff-Bonekaemper 2002) and its impact on social identity and social practice in Berlin during and after the Cold War have also been discussed (Borneman 1992; 1998). In Belfast, peace-lines or 'interface barriers' separate many working class Protestant and Catholic areas of the city. Neil Jarman has described the effect these boundaries have had on the city and its inhabitants (2002: 283–284). He notes how most residential areas have long been dominated by one community, rather than being mixed or balanced. Churches are segregated, as is the school system, work places as well as sports and social clubs. Most people are therefore born, brought up, live, work, socialise and are buried amongst their own community. Jarman goes on:

> The two working-class communities have lived relatively segregated lives since the early expansion of the city in the nineteenth century, but over the thirty years of conflict [approximately 1969–99] these patterns of segregation have hardened … The families who lived in the streets that connected the Catholic Falls Road and the Protestant Shankill Road were in one of the most vulnerable locations and were subjected to extensive rioting and violent intimidation. The already divided communities were further polarized and a 'no-man's land' was established as a boundary zone after people moved away from the interfaces and further into the heart of their community. Initially improvised barricades or rolls of barbed wire segregated the two sides. Soon these were enhanced by more solid sheet steel fences and then further strengthened by a two-tier steel fence so that the barrier reached some 6–7 m in height.

These barriers and physical boundaries have obvious significance in demarcating urban space and defining or imposing a sense of community. Comparisons can be made between the effect of boundaries at different scales, recognising that communities like those in Belfast withdraw from boundaries within cities just as they do from those of nations and states (Wilson and Donnan 1998: 13). The emphasis here however is on exploring that sense of community and cultural identity in those places to which distinct socially or culturally defined groups feel attached (these places are sometimes referred to as 'ethnoscapes'). These are often places where physical constraints are not imposed or constructed, but the interface is more subtle and therefore difficult to trace. In Berlin for example, the Wall is well known. Less well known are the cultural divisions that existed and which began to take shape through the presence of French, American and British service communities during the Cold War period. These cultural differences were certainly apparent before 1989 (Schofield 2003). How far these traces continue to define the character of these areas of the city today, 20 years after the Wall came down, is not known.

Character, Sense of Place and Cultural Diversity

Cultural heritage has over the past decade broadened out significantly to include far more than just the material world – the buildings and monuments with which we are so familiar. Characterisation, as a suite of techniques and ideas for understanding landscape (including townscape) in totality and at a broad scale as a means to promoting informed conservation, is now widely used in the UK and – increasingly

– across Europe (e.g. Fairclough and Rippon 2002). Sense of place forms part of this characterisation agenda, seeking out what matters, why and to whom, based on well-developed principles of human geographical and historic environmental research. In turn, social significance is an important component in determining sense of place (Byrne et al. 2001; King 2003), recognising that values will result from social and personal experience of place, and will therefore in large part be culturally constituted. Modern heritage is also now recognised as having value, however recent or mundane the buildings or places may be (e.g. Jones 2002). And it is also now recognised that cultural heritage can be tangible – in the form of buildings or monuments – or intangible, in the form of customs, language and dialect, musical styles, arts and performance, rituals and so on. All of these are relevant and related considerations in defining and understanding new urban frontiers and the will to belong. A short discussion of these issues will therefore precede some examples.

To begin with characterisation, the 'character' of a place or area is defined by the unique combination of factors and influences ('characteristics') that make it distinctive, and set it apart from its neighbours (Fairclough 2002). Areas of landscape can be distinctive in this way, as can parts of a town. Historic influence is a major determinant of character, alongside its contemporary use and the impact of traditions and customs. Characterisation as an approach to recording these differences seeks to take account of them at a general level, seeing and valuing with a view to managing change. It recognises all areas and their characteristics, not just selected and special areas. Characterisation champions local diversity, recognising the importance of the commonplace and everyday; that these more mundane places are 'recognised by all, creating people's links to history and the past, as well as to identity, sense of place, nature and the future' (ibid.: 30). Characterisation also seeks to engage communities and promote participation, enabling citizens to take part in decision-making about significance and future change. It is also about perception, considered to lie at the heart of understanding. Of particular relevance in assessing the links between cultural diversity and place is the recognition that urban and rural landscapes are a construction of intellect and emotions, containing different ideas, feelings and associations. While heritage professionals undertaking a characterisation exercise will recognise distinctive areas in terms of historic fabric, layout, design and urban topography (an allocentric view, arguably, after Porteous 1996), another level of distinctiveness will involve the perceptions of inhabitants, sometimes of very different social and ethnic origins (a more autocentric view, ibid.). Recognising that degree of diversity and its implications for cultural heritage management and implementing social policy is one of the objectives of this chapter.

Related to character is sense of place and belonging. Sense of place can operate at various levels and with various degrees of connectivity. Culturally, places will have value and significance, and these places may be 'historic' or landmarks in one sense or another, or they may be mundane, such as a street market for example. Either way they may be valued by the community, whose members will have a common cultural link and connection to that place, for example in remembering parades or community events and outings. Families can recognise sense of place, for example in a house they once occupied, or places where family holidays were taken. And sense of place can operate at the personal level, for a whole host of intimate or professional reasons and motivations (e.g. Read 2003, who describes these personal connections with place as sometimes creating enspirited places); place will have significance as home as well as cultural heritage. But significant here is the general connection between memory and place, and the recognition that places are valued at different levels and in different ways dependent upon one's social and cultural context. Early work in this area included – significantly – Dolores Hayden's Power of Place project, which demonstrated the cultural dimension of urban history as relevant to the construction of place and identity (1995). Michael Bell (1997) has written about the 'ghosts of place', a phenomenology of place recognising the presence of those who are not physically there.

Cultural heritage management practice is something traditionally imposed by heritage professionals on communities. Decisions are made by those that are qualified to know, by 'experts' on behalf of their 'constituents'. That situation has changed significantly in recent years, cultural heritage management having become more concerned with participation and accountability. Some of the more innovative work in this area has been in Australia where initial studies with the Aboriginal community on social significance have extended to include work with recent immigrant groups. The Australian Heritage Commission for example have engaged the Chinese community to determine the significance of Chinese heritage places (www.ahc.gov. au/chineseheritage), a guide has been produced to enable migrants to find and document their own heritage places (www.heritage.gov.au/ protecting), while work by the National Parks and Wildlife Service in New South Wales has focused on the Macedonian (Thomas 2001) and Vietnamese communities (Thomas 2002). In the UK, projects have been developed to examine heritage interests amongst the Jewish (e.g. Kushner 1992) and Bengali communities (Gard'ner 2005), for instance.

This is a significant area of development therefore, and one that has enabled sense of place, character and social significance to sit centrally within discussions about cultural heritage planning. These developments have meant that heritage managers and planners can now take account of not only historic fabric and urban and rural morphology, but also the character of places as defined by their communities themselves. In some cases this will be manifest in physical form: religious buildings for example; shops and shop frontages; signage. But ownership may also be expressed (and be potentially recoverable) in less tangible form. As Porteous has explained, experiencing place is multi-sensory, involving smell, touch and sound alongside the obvious significance of sight (1996: 8–9). Intangible characteristics of place can involve the distinctive smells of regional styles of cookery, and the presence of a unique combination of products on market stalls, sounds of music and singing, voices reflecting styles, dialects, languages, and the acting out of customs and traditions. All of these factors contribute to the character of a place, making it distinctive and setting it apart from other neighbouring areas and places.

To summarise then, places will be distinctive and display unique character in a diversity of ways. These will include historic and contemporary fabric, and people's own sense of the past and of belonging, all of which will contribute to that area's sense of place. As we have seen, some such places will be divided by physical boundaries for reasons of security, or for political or ideological reasons of separation. Most cities have distinct communities who maintain this sense of place and separation without hard boundaries. In these cases, the boundaries are intangible, and rely on more subtle forms of enquiry to understand and map their extent and influence. Indeed, amongst some communities, the boundaries will be more a matter of personal experience and an intimate engagement with their community space. It is important to recognise these various manifestations and the means by which boundaries are constructed and experienced, for reasons of strategic and community planning, and for the communities themselves to retain their sense of identity as a contribution to informed conservation and management.

Examples

Tower Hamlets, London, England: Gay Space

This first example concerns methodology, and the means by which communities that are sometimes quite difficult to engage can contribute to discussions of place and identity. It also introduces the view that any will to belong is not confined to groups defined by ethnic codes, and can equally apply to groups defined by social status (e.g. various papers in Pacione 1997), age (e.g. Skelton and Valentine 1998) and sexual preference. Of these perhaps least attention has been given to communities defined by sexual preference, whose sense of place may be governed more by social constraints and sexual practice, and personal safety, than by any other culturally constituted criteria (but cf. Kenney 1998; Moran et al. 2003; Reed 2003).

One particular study by Gavin Brown (2001) of 'Gay men's narratives of pleasure and danger in London's East End' is significant here for two reasons: methodology and meaning. To begin with methodology, the approach taken to assessing sense of place and identity in this area involved cognitive mapping: the mapping by participants of their experiences of place – social; personal or culturally constituted; or experiential. It is a simple method, now much used in geographical studies, and supported by subsequent interviews to interrogate maps and add further depths of meaning. Cognitive maps involve providing maps of the study area to members of the community for them to annotate. The result is layered information that together constitutes a cultural statement of ownership and belonging. It can define boundaries (in this case 'Gay Space' and, within and beyond that, boundaries defining zones of social and cultural significance); it reveals meaning that members of the community attribute to places within those boundaries. Cognitive mapping is a procedure for giving meaning to place, yet at the same time allowing a high level of personal attachment to be expressed in connection with place.

One of the maps described and presented by Gavin Brown highlights some distinct parts of Tower Hamlets: an art studio, and areas of housing for example. The author defined 'gay' areas within the borough as, 'areas that I consider to be "gay". I would probably have to put the proviso on that I consider these areas to be, like, visible middle class, white, gay' (Brown 2001: 53). Another map however shows a very different perception of gay space, its author having drawn a line around the entire Borough; no specific locations are recorded, although many (pubs, parks) were highlighted at interview. A third participant presents the middle ground: some large parts of the Borough as constituting gay space (including the areas of housing noted by the first interviewee), but in this case three large and quite tightly defined 'danger zones' are also identified.

Other communities have no idea of these perceptual boundaries, identified only by members of the gay community, and it may be that only those in touch with gay culture and social practices would recognise that they had indeed entered 'gay space'. It is also the case that perceptions will vary, depending for example on how comfortable members of this community are with their sexuality; to what extent they have 'come out'. These then can be amongst the most intangible of boundaries, defined purely by social practice (but cf. Reed 2003 for an example of gay space defined by a physical boundary). They can nevertheless be defined, mapped and understood as constituents of urban space.

The information contained in cognitive maps, and the understanding they can enable within a wider community, contribute also to promoting cultural diversity agenda and tolerance of difference. As Brown says (2001: 60), this information has the potential to allow discussion and debate, and thus build alliances across these apparent divisions, challenging the continued privatisation of significant (for some) public space. He goes on: 'By listening to ordinary people's descriptions of the urban landscape, radical urban planners and community activists can help bridge the tensions between the right to privacy and access to public space … and attempt to create safer and less alienating cities for everyone.'

Tower Hamlets, London, England: Bengalee Space

This example – which describes character and distinctiveness in London's East End – makes close reference to a study by Jim Gard'ner, being a survey of heritage protection and social inclusion (2005) amongst the Bengalee community – the largest minority group – in the London Borough of Tower Hamlets, where it has developed a distinct cultural and commercial identity within a defined geographical area. This example describes how that identity is defined, rather than focusing on its implications for heritage management and protection.

As Gard'ner explains, this mainly Muslim community has adapted existing religious and secular historic buildings to give them new use, and thereby a new cultural significance. Kristin Posehn's photographs (e.g. Fig. 40) demonstrate how new and existing buildings, colour schemes and street furniture give a physical expression of British Asian culture in streets like Brick Lane, setting this area apart from its neighbours and allowing this appreciation of distinctiveness to feed into management and planning discussions with community workers and leaders. To

Fig. 40 Brick Lane. One of a collection taken by the *photographer* Kristin Posehn for the original publication of this chapter

walk into and out of this part of Tower Hamlets is to clearly move into and then beyond Bengalee space.

Discussions with members and leaders of this community have revealed a diversity of site types that contribute most to the character of the area, and have significance as a result. Religious buildings are prominent, as are community centres, streets or markets, parks and gardens, housing, schools and public monuments and sculpture. The Jagonari Centre on the Whitechapel Road is an example of how importance to the community is recognised and expressed. This community centre was purpose-built for local Asian groups and includes on its street façade mosaics in the Islamic tradition. Community centres like this provide a clear focal point: a venue for community, cultural and musical events as well as services including immigration and general advice, training and employment counselling, and day care for the young and elderly.

Moving beyond the buildings themselves, some streets were identified as having particular significance; Brick Lane for example, which is best known to Londoners for its balti houses and 24-h bagel shops; it is however also the economic, social and cultural hub of the Bengalee community. Here we see the interface between communities most clearly expressed. Brick Lane extends from the south, where Jewish and Bengalee-run clothing manufacturing is based, through the Bengalee retail, cultural and restaurant areas in the centre, to the Pakistani-dominated leather industry to the north. This northern part of Brick Lane houses two Jewish 24-h bagel shops, and terminates at Boundary Estate, now home to many Bengalee families, as evidenced by signage and graffiti tags. As Gard'ner has recorded, Brick Lane now contains more than two dozen curry houses, grocers selling Asian produce and halal meat, Asian video, music and book shops, fabrics and clothing retailers as well as other businesses and professional offices serving the local community (2005).

The significance of Brick Lane has been increased however by the desire of the community since the late 1980s to make a stronger impression on its streetscape; to further impose on an area of mixed cultural affiliations a distinctly Bengalee sense of place and community. As Gard'ner has noted (ibid.), part of the motivation for this was to draw tourists and Londoners to their restaurants and businesses as part of 'Hospitality Bangladesh'. Sense of identity is clearly apparent in the street furniture decorated with traditional motifs and colours (red, green), and street signs in both English and Bengalee. An arch forms a gateway into and out of this cultural space which has sufficient identity (and now also formal recognition) to ensure this first 'Little Bangladesh' can sit alongside the Chinatowns that now form a distinct part of most major cities around the world. Distinctiveness is also apparent in the visual codes of dress, and shop window displays, the smells of food in the market or cooking in restaurants, and in the voices and music streaming from open doorways and windows and market stalls. The sense of place here is all-embracing.

What this example demonstrates is the extent to which urban space can have distinctive character reflecting historic origins and contemporary use. It shows how distinctiveness can indeed be read in the language of graffiti tagging and (literally) of signage, as well as the buildings, the use of space, street furniture and temporary street decorations. But it can also be read through the more subtle traces: the use of that urban space, and its buildings and streets; the exploration of place using all the senses.

Northwest London, England: Jewish Space and the Case of the Eruv

This final example explores the difficulties that can arise when the intangible becomes tangible; when there are proposals to make an invisible imagined boundary 'real' (see Reed 2003 for another example of this situation). In 1991 Barnet Council, north-west London, was asked by a group of Orthodox Jews for planning permission to erect clusters of posts around the six-and-a-half square mile area defining their community (www.NWLondoneruv.org). This *eruv* would take the form of 6–7-m-high posts connected near the top with wire or string creating an enclosure. Not a 'hard' boundary therefore, but something more symbolic and indicative; something that 'exists mainly in the minds of those who believe in it' (Trillin 1994: 50). Most of the boundaries for this eruv were in place already: existing sections of the Northern Line of the London Underground for example, and parts of the M1 Motorway. People who live within eruv in other countries are not even aware of it: in Israel and the United States for example.

The justification or desire for an eruv is worth outlining. Some of the Orthodox observe Sabbath restrictions that include a prohibition against carrying anything unless they are within their own private domain; their home in other words. The purpose of the eruv is to extend that private domain to the boundaries of the eruv (effectively therefore co-operative private territory). Even though this proposal amounted to connecting existing points with wire, and erecting some 85 new posts in an area of London that already had 40–50 thousand posts of one kind or another, it was nevertheless opposed, with anti-eruv groups formed and petitions signed. The case of the eruv attracted the national press. Arguments against the eruv included the escalation of street clutter, but most noted that in an area where people of diverse backgrounds had always lived peacefully together by treating religion as a private matter, it was divisive for a minority to impose its religious symbols on everyone else. 'Where would it all end: totem poles on the heath?' (Trillin 1994: 55–56). The point was also made that eruv poles could provide a magnet for anti-Semitic vandalism, and that the wires would remind refugees of the ghettos and of concentration camps. But interestingly it was quite another view that was expressed most frequently by eruv opponents, that the eruv would create another kind of ghetto, the eruv boundaries signifying an area where Jews live, and where more Jews would come to live in order to avail themselves of the eruv.

Here was a case where one group of residents wanted to impose a physical boundary, for reasons of religious practice, where an imagined (and anticipated) one already stood. The controversy about this is worth recalling, as is the fact that disagreement was also felt within the Jewish community, not just between and outside of it. It took several years for a decision to be reached. The proposal to construct an eruv was finally granted permission six years after community members first went around mapping out its boundaries. It eventually came into use in February 2003.

Interpretations

The three examples cited each contribute separately to interpreting the construction of urban space and identity. They describe the criteria by which identity and character may be considered distinctive, focusing on the tangible and intangible traces of urban life. They demonstrate how boundaries can be clear and marked, or blurred, as in parts of Brick Lane, London. Boundaries can be hard or soft, real or imagined; visible or invisible to the outsider. They can be transient and as ubiquitous as the location of experience; and they can be time specific, applying for example only at certain times of the day, or of the year (Moran et al. 2003). The perception and recognition of boundaries can be culturally constituted and widely recognised within society, or a reflection of personal experience. And one territory may have several boundaries, recognised by different groups within society: the Gay and Bengalee community in Tower Hamlets for instance.

Fig. 41   Nevada Test Site graffiti, on the fence separating it from Peace Camp. *Photo:* author

In the past, boundaries tend to have been recognised where they exist or had existed as hard – often political – barriers. Their significance in other words concerned their physical form and its social and political implications. These are the boundaries of nations and states; the boundaries that have international borders and checkpoints as their contact points; the 'border identities' of Wilson and Donnan's (1998) collection of essays. This collection of work – containing ten anthropological case studies – examined the various ways that international frontiers influence cultural identity. But it also explored the social significance of borders, notably in Borneman's chapter which describes one woman's experiences in pre- and post-unified Germany, and her desire to seek openness and a lack of boundaries through relationships and sexual liberation in her own life. That is really the point of this chapter: that boundaries can have an influence at all levels within society, from the cultural (nation) to the social (an ethnic group for instance) and the individual. Until recently only the first of these – with their easily recognised boundaries and clear cultural implications – were much studied. In the last few years, with the recognition in urban planning and heritage management that boundaries can also exist *and have significance* at a community scale, interest has extended to cover this wider field of view. It is also recognised, and demonstrated here, that these intangible boundaries are identifiable, and will typically define areas that have distinctive characteristics. Finally there is now the realisation that personal space is also something that can be recognised and mapped, marking at the most intimate level experiential space in an urban landscape.

There is also now recognition that community space has equal relevance to those groups on the margins of society. The Gay community in Tower Hamlets is hardly marginal, and neither of course are the Jewish or Bengalee communities. But the voices of youth culture are rarely heard. A project being proposed for Liverpool, to assess the significance of popular music in constructing space and local identity has this as one of its objectives, building on work which has assessed sense of place (the 'hood) and its reflexive relationship with rap music, for example. The homeless, asylum seekers and drug addicts also construct space and create (or live within perceived) boundaries. For example, in Germany, Fiona Smith describes how:

> ... between the GDR revolution and reunification a period of relative openness led, in many areas which had been centres of squatting and alternative scenes, to the establishment of groupings of mostly young people who sought to change the dominant urban planning concepts and to establish alternative housing and culture projects. One such was the Connewitz Alternative in Leipzig, located in a small area of poor quality nineteenth century workers' housing: 'We want to save this unmistakable quarter ... and we want to transform it at the same time and give it an alternative-cultural character' (Smith 1998: 299).

Smith goes on to describe how urban spaces became the domain of largely left-wing youth cultures and those with few political convictions. The right wing scene:

> developed its own geographical strongholds, most often in the modern estates of the GDR period ... One element in the geography of youth cultures then became, very obviously, the divisions and violence between left wing and right wing groups, further increasing the 'youth as problem' discourse and creating particular sites as areas of conflict (ibid.: 301).

It is worth noting here that some of the studies described in this chapter focus on maintaining boundaries and ensuring continuity of character within the urban fabric; some – like these last examples – are more about the removal or softening of such boundaries to improve conditions for those disadvantaged by them.

Meaning

These new urban frontiers may not always be so new, but the emphasis placed upon them, and their significance in understanding sense of community, belonging and the use of urban space is immense. This has relevance both in a contemporary sense and setting, but also over time. Buildings, sculpture and artefacts will remain as tangible links with the community, giving it clear and obvious legibility. Other traces (music, smell, dress and language for example) will have significance in maintaining sense of identity and belonging in the present, but will be given chronological depth only through film, photography and artistic intervention, and the recording of oral historical accounts; there will be no physical expression that can be identified as 'historical'. It is important to recognise that social and political conditions, and planning policy, will cause these urban boundaries to change and reconfigure, and that this process of change and renegotiation is also something of importance, both to the community and to us as archaeologists and historians in charting the evolution of urban space. With increasing emphasis being placed on the recognition of social significance in cultural heritage management and planning, and with a move to improve community participation in the heritage, the full range and diversity of these boundaries, the various forms they can take, and the processes of change, need to be understood.

The three main examples presented here are all from England, where moves to promote cultural diversity and social significance agenda are being encouraged by government, and pursued by national heritage agencies and local authorities. But the significance of new urban frontiers is Europe-wide. Europe's cultural map is extremely fine-grained compared to some other parts of the world, and diversity has contributed in a significant way to the cultural heritage that we – and many visitors from outside of Europe – enjoy. This sense of place and identity, of community and belonging, and of the transmission of meaning and significance over time, has value therefore for maintaining that diversity and ensuring communities retain their distinct sense of place and of their own pasts for the benefit of all.

Note

The original chapter contained a photo essay by Kristin Posehn. These photographs were made within the Bangladeshi community of Brick Lane and Tower Hamlets, East London, as described in the second case study of this chapter. Kristin writes: 'In making these photographs I was encountering these areas for the first time. I was drawn from the start to the physical and spatial manifestations of the community's boundaries; in retrospect, to photograph was to discern and make clear boundaries that were otherwise only intuitively sensed. In these photographs, gates, doors, windows and intersections function as membranes between the more intangible qualities of the community's inside and outside'. It is recommended that readers view these images in the original publication.

Chapter 14
Constructing Place: When Artists and Archaeologists Meet

Art and archaeological practice are closer than some might think. Some artists work with archaeological material, and interpret archaeological sites through a diversity of approaches and media. Equally, art can become archaeology – Francis Bacon's studio was 'excavated' after his death, for example. Even the processes overlap: archaeological fieldwork as performance; the similarity of 'incavation', intervention and excavation. Here it is argued that the role of the archaeologist, indeed the very definition of archaeology – to characterise and contextualise material records of the past – can usefully be expanded sometimes to include the contributions of artists.

Overturning Convention

Today we are all archaeologists. (Michael Shanks, commenting on Holtorf 2005).

Convention limits archaeology to the study of material remains from the remote past – from antiquity. Recently this definition has been expanded to include contemporary archaeology, which takes the definition of archaeological practice and theory to a logical next stage: the archaeology of us. Buchli and Lucas (2001) talk about contemporary archaeology challenging the 'taken for granteds' of modern life; while Graves Brown (2000) speaks of archaeological practice and theories serving as a critique of the world we ourselves have created. Interestingly, in 1966, an Institute of Contemporary Archaeology was established by the Boyle Family, a family of collaborative artists based in London, to give context and identity to their work 'Dig' (www.boylefamily.co.uk), albeit as a 'light-heated institution with no particular membership' (Elliott 2003: 15–16). As Sebastian Boyle told me: 'it fitted in with our approach of trying to be objective, to see the world as it is, accepting reality and not trying to embellish it for the sake of art'.

Coincident with the emergence of this broader definition of archaeology, archaeologists have become increasingly trans-disciplinary in their approach towards material culture. The limited attention paid to artistic practice and archaeology however is surprising, notwithstanding Renfrew's (2003) wide-ranging and influential study of modern art and archaeology, work by artists such as Anne and

J. Schofield, *Aftermath: Readings in the Archaeology of Recent Conflict*,
DOI: 10.1007/978-0-387-88521-6_15, © Springer Science + Business Media, LLC 2009

Patrick Poirier who have been doing art about archaeology and art history for over three decades, and the active interest of Finn (2004, and personal communication), Holtorf (2005), Jameson et al. (2003) and others. Focusing on conflict heritage (Schofield 2005a), this essay will review instances of artists working with, and providing interpretations of, contemporary archaeological sites to demonstrate how these different approaches, taken together, can build understanding of the world around us. Three types of application will be considered:

1. Art as an archaeological record; the idea that we create as well as consume material culture, and the past as a renewable resource
2. Archaeological investigation as performance
3. Art as interpretation, as narrative and as characterisation

In each application, the close similarity between art and archaeological practice is emphasised. For example, Bourriaud (2002, cited in www.gairspace.org.uk/htm/bourr.htm) defines art as an activity consisting in producing relationships with the world with the help of signs, forms, actions and objects, and refers to the contemporary artist as a semionaut, inventing trajectories between signs. Both statements are equally true of archaeology. Equally, what is significant in archaeology is the process of doing it, more so than the results of the endeavour. To cite Bourriaud once more: 'present day art does not present the outcome of a labour. It is the labour itself, or the labour to be'.

Regarding context, a significant British artistic movement in recent years concerned the Arts Placement Group (APG) (Broekman and Berry 2002). Emerging in London, APG's recognition of social context and the merits of conceptual art informs many artists operating today outside of gallery spaces, in an expanding and important field where dialogue and process are dominant; where the function of art is decision-making. The APG's view that 'context is half the work' applies to numerous of the examples that follow, demonstrating the depth of the Group's influence (http://www.tate.org.uk/learning/artistsinfocus/apg/chronology.htm). Needless to say, context is fundamental to the archaeological process and to reading and interpreting material culture.

The essay concludes that archaeologists and heritage managers can usefully include artistic work as a legitimate and constructive means of interpreting these more contemporary heritage places, and further that the recording of such places by artists can capture their character – their aura – better than any conventional record produced by archaeologists or historians. The reason for this:

The world of the visual arts today is made up of tens of thousands of individuals, most of them doing their own thing. Among them are creative thinkers and workers who are nibbling away all the time at what we think we know about the world, at our assumptions, at our preconceptions. Moreover, the insights that they offer are not in the form of words, of long and heavy texts. They come to us through the eyes, and sometimes the other senses, offering us direct perceptions from which we may sometimes come to share their insights. The visual explorations … offer a fundamental resource for anyone who wants to make … sense of the world. … It is not that this resource offers new answers, or that it will directly tell us how we should understand the world. On the contrary, it offers us new, often paradoxical experiences, which show us how we have understood, or only imperfectly mastered, what we think we know (Renfrew 2003: 7–8).

Furthermore,

[images, like] no other kind of relic or text from the past can offer ... a direct testimony about the world which surrounded other people at other times ... [T]he more imaginative the work, the more profoundly it allows us to share the artist's experience of the visible (Berger 197: 10).

Art as Archaeological Record

Artistic intervention is one small but significant dimension to an ever-expanding archaeological record. I recently argued for the recognition of modern graffiti as archaeological evidence for example, intimating the character of urban space, giving voice to subcultures within urban communities and their resistance to gentrification and globalisation (Schofield 2005b; see also www.grafarc.org for graffiti archaeology images and reference to the graffiti artist as 'archaeologist'). War art is another example (e.g. Cocroft et al. 2006), as is 'Land Art' which has particular relevance given its obvious reference to earlier archaeological sites and late twentieth-century monumental architecture. Once these new places and things are created, their creation is in the past, and thus archaeological, at least by anything other than the most conventional and literal of definitions (e.g. www.changeandcreation.org and Bradley et al. 2004).

Indicative of this broader definition of archaeology was Holtorf's (2004b) incavation in 2001, in which eight contemporary and mundane domestic assemblages were buried in the garden of a Berlin townhouse. As Holtorf explains:

whether one incavates or excavates, archaeologists ... construct the past and its remains like artisans create their craft. It takes desire, creativity, skill and persistence. ... Incavating is not however about faking archaeological evidence, about making archaeology appear a futile exercise or about drastically diminishing the cultural impact of what is being hidden in the ground. Instead, what is incavated is archaeological evidence in itself ... (2004b: 47–48).

Artists working with aspects of conflict have also contributed to the archaeological record in this way, by creating material records. In 1999 and 2001 I visited District Six, Cape Town, an area of the city where a long-established mixed race community was forcibly removed under the apartheid regime's *Group Areas Act* (see Chap. 2). Following collapse of the apartheid regime, District Six became a focus for its former residents, one aspect of which – in 1997 – was a public art event that sought its reclamation. Artists with connections to the District were commissioned to produce installations and artwork that reflected the history of the place, their experience and attachment to it. The archaeology of some of these interventions remained when I visited in 2001. Bedford and Murinik review the works and give them context:

Through their various works, artists drew our attention to District Six as place, a physical landscape once densely populated and now scarred and barren, but as metaphor for a range of displacements. The wholeness of the place and the totality of its meanings were vividly contrasted with the lost and the broken: fragments indicating the break-up and fracturing of society and the loss of things precious to the soul. The project should be approached in a similar way; understanding it as an attempt by a group of artists to gather the many fragments, both physical and narrative, that commemorate both an era and its people (nd: 13–14).

Roderick Sauls' *Moettie My Vi'giettie* harnessed the incessant wind that residents remember. It recalled the colour, textures and movement of carnival, and reflected on the dispersal of people. A frame with cloth fragments represented these characteristics and made historical references to place and history. Andrew Porter's work (untitled) compares closely with Holtorf's incavation (ibid.), being a completed excavation site which required backfilling, a simple mechanical task on most excavations, but heavy with symbolism in this instance. The author described the intention as,

> getting the viewer to place the excavated soil back into the hole from which it was taken. I wanted this participation to function as a kind of ritual, the physical nature of which would encourage the viewer to think about District Six and on another level to lay the soil to rest.

Away from District Six, on a more intimate level, and as a means to interrogate personal and collective relationships to South African British colonial history and its current personal and public residues of identity construction within the context of postcolonial, post-apartheid South Africa, the artist Leora Farber uses her skin as a 'figurative and metaphorical site of intervention, for the grafting of tensions, ambiguities and differences' (Farber 2005: 301–302).

A series of photographs show a woman in Victorian/African-style clothing, seated variously in a formal garden, an aloe grove, and in the bush. Throughout the sequence, one sees a 'women's work' turned inwards upon itself, with the woman appearing to work her skin rather than fabric, thus negotiating the sense of being British in an African/post-colonial context. The woman is seen to sew into her forearm a seedling aloe plant, which grows to arm's length through the sequence of images.

> The gentle politeness of the needlecraft action, executed in the pleasantries of my … surroundings, is undermined by the horror of self-violation (Farber 2005: 306).

Like backfilling earth in District Six, intervention here represents a negotiation of space and identity, though in this case the incavation is more intimate … more shocking.

In Berlin, artistic installation and intervention since 1989 follows earlier traditions of decorating the West-face of the Berlin Wall. As Feversham and Schmidt (1999: 154) put it: 'Whilst the Wall stood, it served as a work of public art, a blank slate for the expression of private and public resistance'. And it was these interventions that assured the preservation of wall fragments and sections, the Museum of Modern Art in New York for example, buying a section on its artistic – as opposed to its historic – merit.

As a result of the creative energies that the Wall inspired, the artistic tradition has continued. In 1990, the East Side Gallery was created, adopting the previously pristine East-face of the Wall as a canvas. Some 118 artists from 21 countries produced paintings here, and for artists from the East in particular the experience was a profound one. Other post-Cold War works in Berlin include Frank Thiel's tall steel pillars at Checkpoint Charlie, displaying luminous colour photographs of American and Soviet soldiers facing East and West respectively, and *Rabbit Sign* by Karla Sachse. Feversham and Schmidt explain:

> 120 life-size silhouettes of rabbits cut from sheet brass have been stuck onto the surface of streets and pavements on the site of [a] former crossing station. ... It is interesting to recall that an East German children's club run by an artist friend of Sachse's used to hold a 'rabbit party', and raise the 'rabbit flag' – an oblique reference to the idiomatic German expression '*das Hasenpanier ergreifen*' (i.e. showing the rabbit flag = running away), an expression so archaic that the Stasi failed to grasp its subversive undertones (1999: 156).

At Peace Camp (Nevada), a camp site occupied periodically outside and in opposition to the Test Site there, residents created landscape art as expressions of opposition. As we have seen (Chap. 5), these works are numerous and diverse in form, but all are representative of the communities within which they originate. Here art forms the basis for archaeological interpretation of the site, its occupants and the activities conducted there. This is a remote and harsh desert environment where stones and rocks are the only materials with which to work; and they have been used to remarkable effect, the shapes, sizes and colours of stones giving texture to artistic representation. In recording these contemporary remains, the diversity of religious and philosophical backgrounds is obvious in the diversity of representations: Christians (fish and chi-rho symbols); Franciscans (Franciscan crosses); New Age religions (offerings of various kinds); traditional owners the Western Shoshone (tortoise and snake representations); as well as Buddhists, Hiroshima veterans, Jews and so on. This is landscape art, reminiscent of Richard Long, Michael Heizer and others, demonstrating in an obvious and literal way how art can be archaeology, and archaeology art. The connection with Michael Heizer's work may be particularly significant here, given its close geographical and thematic proximity to Peace Camp. Heizer's 'Land Art' is substantial. In 'Complex One' (1972) for example, in Hiko, near Nevada, an enormous pile of earth was sandwiched between two reinforced concrete triangles with large concrete beams inflecting the structure. Robert Hughes described 'Complex One' as recalling,

> the enigmatic structures left behind by America's various nuclear and space programmes, which by the 1970s were already beginning to seem an archaeology of the age of paranoia (www.articons.co.uk/heizer.htm).

That some reference is made by Heizer to the concrete and monumental architecture of the test site seems obvious. It also seems unlikely, given the timing of the formation of Peace Camp, and 'Land Art's connection to hippy culture in its move away from developing technology to embrace beauty and nature' (ibid.), that the protestors' landscape art did not also in some way hark back to the test site, filtered through their knowledge and familiarity with the emergence of Heizer's Land Art.

Archaeology as Performance

Increasingly, archaeology is seen as a performance (Pearson and Shanks 2001;
Goldberg 1979), figures in a landscape, doing archaeological work in a conventional sense, but as actors in a wider study of people's interaction with place.
Performativity is an aspect of this, noting that human activity can be passive
while the space in which activity occurs is active. Space performs us in other
words, not the other way round. Paul Virilio's (1994) seminal study of bunkers of
the German 'Atlantic Wall' could almost be considered in these terms. His exploration was both a personal voyage of discovery, as well as an interaction with
people and with place. It was also an extremely good field survey which, like the
Boyle Family's Institute, did for archaeology what we were later to do for
ourselves.

Performativity is evident also in the excavation conducted at Upper Heyford in
1997 (*minus-F-one-eleven*). This was a small-scale inhabitation of a Cold War
airfield, with a view to connecting out to the spaces of the surrounding landscape
(Fig. 42). Part of the study involved creating an excavation trench, neatly ordered

Fig. 42 minus-F-one-eleven – ex-US Air Force base, Upper Heyford, Oxfordshire – trial pit.
Photograph courtesy of Peter Beard

spoil heaps on the former runway and the symmetrical alignment of turves. Here, the process of archaeological investigation is presented as performance, by an artist (Peter Beard: personal communication) replicating the archaeologist.

Artists also record archaeological practice as art. Louise K. Wilson (2003, 2006a) has done this at Spadeadam (Cumbria), a Cold War missile testing site. In a recent discussion of her collaboration with archaeologists studying the site, she said that:

> As an outsider to the means and processes of archaeological surveying, it was becoming interesting to read what the archaeologists were doing as some strange form of perform-ance or ritual. In order to take GPS measurements, they were physically traversing the sites – climbing over and around the disintegrating concrete. There was of course something very ironic about the use of sophisticated GPS kit to survey the test stand for a doomed satellite launcher.
>
> …
>
> There has to be an intimacy between the physical act of surveying and the architecture to get the fullest story possible. All sorts of remote archaeology need this intuitive layer. Someone is needed on the ground if there is to be a real concern with accuracy.

The research practices and emotional investment amongst artists and archaeologists are thus closely matched: the attachment to place, the landscape and its physical properties, and the act of 'doing' something on or with that place to create a narrative or interpretation. In the context of performance, art and archaeology can perhaps only be separated by their ultimate purpose, and where those purposes are diverse and overlap, so the boundaries between artistic and archaeological endeavour can be blurred to the point of becoming meaningless. In all of these examples, the result is artistic, and it is archaeological, being a record and an interpretation of the material evidence.

Performance can also have a more literal meaning. Lucy Orta's work for example (Pinto et al. 2003) is often a critical response to sensitive areas of soci-ety, reflecting on themes such as social inclusion and community, dwellings and mobility, and recycling; making the invisible visible. *All in One Basket* and *Hortirecyling* (1997–99) grew from Orta's shock at witnessing food wasted in markets. Her response was to organise the collection of leftover food and ask a celebrity chef to cook it, resulting in a buffet for passers by, uniting rich and poor in a demonstration of gastronomic recycling. A further work was a silent peace protest, and an intense political statement. *Transgressing Fashion* involved models (myself included) wearing army uniform and gold leaf, and passing through London's Victoria and Albert Museum, to mark the handover of Iraq in June 2004 (www.showstudio.com/projects/transgress/start.html). This isn't archaeological in the same material sense, but it is an interpretation of the contemporary world, of the material objects that characterise it, and of people's interaction with both. Here, more than in most of the projects described, we occupy the ambiguous middle ground: the space where artists, anthropologists, philosophers and social commentators meet, and which archaeologists occasionally (and increasingly now) visit.

Art as Interpretation

Contemporary art is being used increasingly to interpret historic places (e.g. Parr 2006; Hayden 1995: 67 ff). In the United Kingdom, English Heritage has worked with the Arts Council and Commissions East on a 'Contemporary Art in Historic Places' initiative including interventions by artists at places as diverse as Felbrigg Hall (Norfolk), and Orford Ness (Suffolk – see below). In the United States, a joint initiative by Boston National Historical Park and the Institute of Contemporary Art has involved New England artists re-interpreting the city's historic fabric. Artist Krzystof Wodiczko for instance interviewed the mothers of murder victims in Charlestown, which had a high rate of unsolved homicides in a neighbourhood where the code of silence ensured that no one would be held accountable. By night, Wodiczko projected a film of the interviews on Bunker Hill monument. Arts critics raved, and some residents raged. But as the artist said, 'Let the monument speak' (Anonymous 2005: 4).

My main contention here is the view that art can provide a significant new dimension to the understanding and interpretation of place. As Feversham and Schmidt (1999: 166) have said, 'There is an argument that contemporary art has a vital, if largely unsung part to play … acting as an agent provocateur in re-energising spaces which by virtue of their very historicity are in danger of being perceived as sacrosanct'. Thus Stefan Gec's proposal for the fourth plinth in Trafalgar Square – 'Mannequin', being two wooden life-size replicas of sea-launched Tomahawk cruise missiles – sought to re-energise and re-politicise the space, 'exploring the concept of victory and its commemoration in the twenty-first century' (from www. bbc.co.uk/london/yourlondon/fourth_plinth_gec.shtml. Accessed 1 September 2005; for other work by Gec cf. Patrizio 2002 and www.stefangec.com).

Along similar lines, Gair Dunlop's *Vulcan* (http://www.gairspace.org.uk/vulcan/htm/vulcintro.html) describes the contradiction between the idyll of the English country house, and the impact of militarisation and new technologies upon it (Fig. 43). It is a work that considers the transformation of our awareness of overlapping structures and networks in the countryside. It links the defensive with the decorative, and with the transience of the structures of militarism. A statement on the project Website notes for example that:

> if the landscaped garden can be said to embody the Picturesque, then the always alert nuclear bomber embodies Sublime Terror.

This purpose of art for interpretation has particular relevance in its association with heritage management practices, and especially the emerging focus on landscape character, being the qualities that make local places distinctive (e.g. Fairclough 2003). Art has for centuries sought to represent the characteristics of place (de Botton 2003, Chap. VII; Howard 1991). Some examples follow that illustrate this potential through the work of contemporary artists exploring the character of former military sites.

Since the end of the Cold War, the defence estate, in the United Kingdom and overseas, has reduced in size, closing down military establishments and bases that were surplus to requirements, and finding new uses for them. Some of these places

Fig. 43 Sublime, melancholic. Vulcan parchmark in a Norfolk garden. *Photograph* courtesy of Gair Dunlop

are iconic in people's experience and memory of the Cold War. Greenham Common is a notable example, where cruise missiles housed in shelters became a focus for opposition, putting Greenham in the front line socially and politically. At the time of closure, the owners of the site commissioned a photographic artist to record the process of drawdown, change and re-emergence (Kippin 2001). Greenham was also part of the inspiration for the Wilson Twins' Turner-prize-nominated *Stasi City, Gamma, Parliament, Las Vegas, Graveyard Time* (Corin 1999).

Other redundant (or deactivated) Cold War sites in England have been the subject of attention by photographers and film makers (e.g. Watson 2004, 2007), as have sites in the former Soviet Union (e.g. Honnef et al. 1998). Angus Boulton's film *Cood bay Forst Zinna* (2001, 2007) perfectly captures the character of an abandoned camp in the former East Germany, occupied by a conscript army far from home. The instructional drawings (to cater for the diversity of languages amongst the soldiers), the sports facilities, and the obvious speed with which the place was finally abandoned, are all represented in this film and captured in a way that a conventional archaeological record could not have achieved. Indeed *Cood bay* exemplifies film as a significant and compelling means of expressing character and conveying to viewers the power and meaning of place. I am reminded of a recent review of Vancouver artist Stan Douglas's work in *The Guardian* newspaper. It said:

Projected images have a particular capacity to reach into us. They may be insubstantial creatures of light and darkness, but that's how they worm their way in. We replay memories as though they were our own home movies. And other people's movies, and other people's stories, become by some circuitous route, our own. The events unfolding up on the screen may not have happened to us, but the movies did. And now movies are in us, and TV is in us, and our relationship to them is no longer simply as witnesses and viewers of once-novel media. They frame our dreams and, in some part, our waking lives.

Like film, photography can also capture the visual characteristics of place. Seawright's (2003) photographic record of post-war Afghanistan, according to John Stathatos (2003: unpag.) echoes the 'occasionally eldritch quality of Afghan landscape', and describes the paradox of this landscape, 'which always seems to be concealing something'. He goes on:

> The colours are generally muted, greys and light browns, mineral purples and ochres; even the rare greens seem faded. Above all, whether in the mountains or the desert, very little seems to obtrude on the landscape, which is made up of foreground and background, but only rarely of middle ground; when something does appear in the middle distance (a rider, a tree, a ruined tower or wrecked vehicle), it does so with unexpected presence.

As part of the same commission, Langlands and Bell visited Afghanistan in October 2002 (Langlands and Bell 2004). Again their work reflects on the character of a country at war, picking out details of landscapes where little was hidden from view. They visited and photographed the main American airbase at Bagram, the site of the destroyed Buddhas at Bamyan, the Supreme Court in Khabul, and – now famously – the former house of Osama Bin Laden, an aspect of the commission that Angela Weight describes as a 'curiously transgressive work: the war on terror meets Grand Designs' (2004: 286). She goes on to recognise this as both, 'a work of art and an extraordinary historical document' (ibid.: 285). Langlands and Bell themselves went further, recognising character as a central feature of their work:

> Architecture is one of the most tangible records of the way we live. Buildings tend to encapsulate our hopes and fears at many levels while also reflecting the persistent human will to plan events. This is evident whether we are considering the monumental edifices of the twin towers in New York, or this modest group of structures at Daruntah. In both contexts we can discover a language of intentions in the character and fabric of the structure (ibid.: 221).

Sound is a further dimension of character (Porteous and Mastin 1985), whether the sounds of the place itself, unfiltered and raw (e.g. Mills 2005), the creation of a soundtrack based on those distinct auditory characteristics (e.g. zoviet*france's *Tramway* project, 2000; and their soundrack accompaniment to *Spadeadam*, 2003; July Skies' *The English Cold*, commemorating the Air Force's presence in East Anglia, 1939–45; and Yannis Kyriakides' *a conSPIracy cantata*, 2001, and *Buffer Zone*) or the performance of work in place, for the very specific combination of effects it can have on people's perception of it. An example of this last category is Louise K. Wilson's *Orford Ness: A Record of Fear*, in which she invited singers to perform madrigals in some of the Cold War test cells (2006b).

One can also directly engage with places, testing their qualities, enduring them and thus attempting to understand their impact on perception and behaviour.

Stephen Turner recently spent a month, unaccompanied, on an abandoned and isolated sea-fort in the Thames Estuary (Turner 2006). The building was constructed during the Second World War as part of the anti-aircraft defences for London, and was later re-used as the home of pirate radio stations in the 1970s, since when it has been abandoned. The experience of living alone here forms part of the artist's involvement with odd and abandoned places, places where he can note at first hand changes in the complex relations between the natural environment and those who inhabit it. Turner's work concerns aspects of time and the dialectics of transience and permanence. These issues were reflected in the artist's daily Blog, and his photographs of the detailed and intricate archaeology of the place so few have visited since it went out of use.

Similarly, Neville Gabie explored and interpreted the remote coastal landscape of a military testing establishment at Foulness, recording the problems of access, and the reaction of others to his project. The publication *Coast* (Gabie 2005) includes kite-borne video films, and records a collaboration with an African writer, trawling local car boot sales, and a journey on a Russian cargo ship (www. coastart.org).

Experience is a central part of The Arts Catalysts' *Zerogravity*, in which 20 or so artists created works in zero-gravity environments including at Star City, the cosmonaut training centre near Moscow (Triscott and La Frenais 2005). A more literal 're-enactment' forms the basis of Jeremy Deller's (nd) *The English Civil War Part II*, in which a full re-enactment of the 1984/5 Battle of Orgreave was staged, with former miners and policemen taking each others' roles.

Conclusion

Using conflict as my example, I have used this chapter to demonstrate that contemporary artists and archaeologists are not so far apart in their approach to recording and understanding the world around them, a point indicative of the increasingly post-disciplinary world in which all professions and practitioners appear now to operate. The examples make the point that these overlaps and correspondences create an innovative and effective methodology for interpreting dissonant heritage; they bring the materiality of conflict, its visual and auditory signatures and signifiers, to a wider and more diverse audience than otherwise would be the case, and challenge that audience in new and provocative (sometimes shocking) ways. Importantly though, and this comes out most strongly in the third category presented here – art as interpretation – artists may be better able to capture and document the contemporary *character* of these places of conflict (their Zietgeist) than archaeologists and historic geographers could ever achieve. This is because they share with archaeologists an acceptance of reality combined with an eye for detail, but examined and represented through the developed senses their training, experience and instinct provide. Of course geographers, archaeologists and heritage professionals will have a significant role to play, in map regression for example, and understanding longer-term patterns of change, but artists may be better able to capture the essence of the place, and people's contemporary perceptions of and interactions with it. As the film maker Dziga Vertov said in 1923:

> I'm an eye. A mechanical eye. I, the machine, show you a world the way only I can see it. I free myself for today and forever from human immobility ... Freed from the boundaries of time and space, I co-ordinate any and all points of the universe, wherever I want them to be. My way leads towards the creation of a fresh perception of the world. Thus I explain in a new way the world unknown to you.

Art is subjective, and individual, and may be it is this very subjectivity and individuality that gives artists the freedom to capture the character of place in the way it does.

Ultimately, what I am suggesting is a form of consilience, defined by the biologist E.O. Wilson (2003) as the pooling of experience, knowledge and methodologies, to gain a rounder, more holistic view of the subject. Art and archaeology can become much closer than they are presently, both as research practices and for experiencing, interpreting and theorising the contemporary past, pooling memory and materiality to create new and previously unforeseen views of the familiar world around us. As Feversham and Schmidt so eloquently put it:

> Contemporary art – vital, provocative, of the moment – when forming a partnership with an historic building or place can act as a conduit to the interchange of time, memory and present history, challenging and de-naturalizing complacent assumptions, establishing a building in the public consciousness and investing it with contemporary relevance. This certainly constitutes a valid and powerful facet of conservation which transcends conventional preservation techniques, simultaneously stimulating debate and working with change rather than striving for immutability (1999: 166).

Afterword

Ghosts

'Berlin is a haunted city', Brian Ladd said (1997 : 1), before noting and describing how buildings have so many stories to tell. These stories concern famous leaders of high politics and high culture, but they also concern the lives of ordinary people as well. And that's what this short essay is about. Nothing dramatic; nothing hugely significant in the grand narrative of Berlin's traumatic and troubled past. It is an essay about what at the time seemed ordinary and everyday, but has taken on extraordinary significance in terms of my professional career, and – specifically here – the content and context of this book. But before reflecting a little further on this, I should tell the story.

+++

Memories fade, and sometimes we remember things through a complex filter in which photographs, stories, subsequent experiences and – well age – can reshape past events, giving them a new focus that reveals things one had not noticed previously, or allows them to attain new significance. Loss of detail is inevitable with the passing of time. Sometimes events remain clear, particularly those with significant and vivid impact. However, more often the picture we paint now will have been changed in composition or content; it will be hazier and more prone to exaggeration and reinvention.

As a child, living in Berlin, I had an experience which I am now convinced changed my life. I don't believe it is an exaggeration to say this. I also do not believe it is unscientific or unprofessional, overdramatic even, to talk about such a change in this way. I have never held-back from describing this event, with friends, over a drink. But I have always been reluctant to go further, to commit it to print for example. Maybe I was reluctant to take this to a level beyond the ordinariness that it remains for me. It was not a big, earth-changing event – something deserving of a sharp, collective intake of breath or demanding the consolation and comfort of others. It was at best something small, local; something that for most people is an item of passing interest, quickly forgotten; at worst it was simply insignificant and unbelievable. To describe something as life changing, even at this personal level, is a grand claim, but I can make it here because it is the context that this book provides where that change is most directly felt. My 'experience' gave me insight and a perspective on the past that I have not appreciated until more recently. So here it is: in 1973, in the attic of the Officers' Mess on RAF Gatow (Berlin) I saw a ghost. I was not the only one – there were four of us, all of the same age and we all saw it. Clear as day, for 4–5 seconds. In a building designed and constructed for military use sometime early in the twentieth century, in a part of the building that had been the second floor, but which the RAF chose not to occupy, we saw a nurse going about her business. An 'in-between' moment, in-between worlds; between consciousness and the sub-consciousness; a moment in the past–present. Something to tell my friends, and something to tell you, now.

This is how I remember it now, thirty-five years later, unembellished and devoid of some details I suspect were added later. This is what happened in its raw, unfiltered state, so far as that is possible to describe after a third of a century.

I shared much of my time then with three friends. All were service children, whose parents were officers in the armed forces. Gavin and Alison, like me, came from RAF families. Virginia was slightly older – 13 maybe – but her maturity exceeded that by a year or two. She was definitely the boss! Virginia's father had an interesting job. He was an army officer, and had the responsibility to fly, in a British army helicopter, the full length of the Berlin Wall every day, as a presence presumably, and perhaps to make a record of any changes he witnessed, or of anything unusual or noteworthy. After retirement I encountered him again at the Museum of Army Flying at Middle Wallop where he was curator. Because of his job, and the fact that the helicopter flew from Gatow, Virginia and her family lived on the RAF station, and in fact had the end of one of the wings of the officer's mess. I was a regular visitor there. She had the school hamster one weekend and I spent the best part of a day trying to help her retrieve it from behind the bath! Otherwise life passed largely without incident.

Virginia's home occupied all four floors of the mess, comprising a cellar, ground floor, first floor and second floor attic. The ceilings of the two main floors seemed unusually high I recall. The mess was a building of two parallel wings, with a section joining the two at one end where the public rooms were concentrated – bars, ballroom, dining room and so on. The television room was here, where I watched much of the 1972 Munich Olympic Games. Outside, and backing onto our back garden were the tennis courts, of which no trace remained when I revisited the place in 2003. Virginia lived at the opposite end of the building, at the end of one of the wings. An open stairwell ran from top to bottom here, from cellar to attic, linking the floors. I once startled a cat on reaching the top of the staircase, and watched it jump, falling the full distance before running away, apparently unharmed.

The four of us spent a lot of time at Virginia's house. It was central to where we all lived, but more importantly Virginia's attic had been occupied by her and so become our own place – a place we called 'home', in a child-like, make-believe sort of way. 'Home' was a single room at the very end of the wing, with a window in the gable end. Facing this was a door, which reached out into what seemed to us the massive expanse of the attic, in which those that managed the mess had stored all sorts of stuff over the years, including some wonderful old furniture, crockery, pictures and so on. Perhaps because of our fathers' respective positions, and perhaps because no-one really cared, we had permission to take what we wanted from here, and had, over time, furnished our home with soft chairs, tables, carpets and framed prints. The room was a wonderful space and we loved spending time there. Whether this had anything at all to do with a sense of history, with the building's depth of occupation and use, I don't know. I like to think it did, but it seems rather unlikely. I was far more concerned in those days with whether Peter Lorimer had scored for Leeds, or whether Slade's latest offering was as good as the last. My school history report at this time stated: 'This boy has no interest in the past whatsoever!' It seems unlikely therefore I would have taken such a thoughtful interest in my surroundings.

The view through the door into the attic is one I clearly recall. I was reminded of it recently whilst watching a 1966 black-and-white production of *Alice's Adventures in Wonderland*, directed and produced by Jonathan Miller and starring Peter Cook, Peter Sellers, Sir Michael Redgrave, Sir John Gielgud and a whole host of other well-known serious and comic actors of the time. The film was made partly in and around the vast abandoned military hospital at Netley (Hampshire), opened by Queen Victoria and dramatically sited on the eastern shore of the Solent. This was an almost unimaginably huge building, with corridors close to half a mile long, along which American troops are rumoured to have driven their jeeps in 1945 (Hoare 2001). The warren into which Alice disappears is the empty hospital, and there is a scene of her running along the corridor, with the curtains billowing out from the open windows. The light, the depth of view, the breeze and Alice's anxiety make this a memorable scene. In all of these characteristics, ours was a comparable experience. Our view was also of a seemingly endless corridor, strongly symmetrical, bathed in light from the dormer windows that existed one to a room in all the rooms placed evenly and opposite each other as far into the half-light as one could see. The combination of forgotten, discarded and unused 'stuff' and draughts throughout the attic kept the configurations of dust particles constantly afloat, changing, moving on the air between each pair of doorways ... the effect of a lava lamp almost, for those that enjoy their magical and hypnotic effects. Looking back, the place did seem almost magical. The sort of place, you might say, where one is more likely to *see* things, and I have sometimes wondered how much this might all have been imagined, a story manufactured in the subconscious and enabled by the otherworldliness of the place we spent so much of our time. But in this historic military building on this particular afternoon I did see something, someone; of that I have no doubt at all. What is more, we all saw her, as plainly as we saw each other.

One thing lacking from our home was a large table – the sort of thing one might use as a dining table. Don't ask me why we felt the need for such a thing – perhaps because it was there! But we were told that in the far wing was a room of tables, and we could take our pick. So, we opened our door and headed off up the corridor, turning right at the far end, then right again into the parallel wing, to a room where tables were indeed stored in abundance. We took one and headed back, one of us on each corner. I recall stopping occasionally – there was some issue about who felt more comfortable at the front and at the back; there was a bit of swapping around. As we approached our room I was on one of the front corners, with Gavin I think, the girls on the back. And then, as though it was entirely normal and routine, and we were moving a piece of furniture along a hospital corridor, a nurse passed directly in front of us, out of one doorway and into another, and disappeared. Facing forwards as I walked I had a very close view. She glided, and was a faded, ethereal presence. We saw through her, as is typical in descriptions of ghostly encounters. She had a folded head-piece, a long skirt, and was carrying a small tray containing objects. This is what I remember now. A more detailed description would no doubt have been possible back then. Perhaps a photo-fit of the uniform could have dated her precisely? She stared ahead – no sideways glances; no sense of acknowledgement. We had stopped, hesitated for a moment after she vanished, and then hurried on – with our table – to the room, closing the door behind us. One response might have been to sit down and discuss amongst ourselves what we had seen – we all knew that we had all been witnesses; it was obvious from our collective reaction. But instead, on Virginia's initiative, we headed straight downstairs to where her father had recently returned from work. I recall being in her sitting room, and together describing, in chaotic fashion, what we had all just seen. There was no doubt about what we had seen or, importantly, that we had all seen the same thing. We were reassured by each other's corroboration. I then went home and told my own parents. To their great credit, and to the credit of Virginia's parents, they never doubted us or what we had witnessed.

We did spend more time in our room, and in the attic, but never saw anything like this again, and I have never had any other comparable experiences. Since leaving Berlin aged 11, I have had no further contact with Gavin, Alison or Virginia. It would be interesting to hear their accounts of this now. Maybe their lives and their line of work do not maintain the same close connection to this experience, or perhaps they have forgotten about it. I doubt it though. Some things are hard to forget so completely.

My father did some research, amounting to asking around at the mess, and returned with the news that the building had indeed been used as a hospital at some stage, presumably during the Second World War. Despite attempts to confirm this, I have been unable to find any information relating to this building, though I continue to search when the opportunity presents itself. I did look up Second World War nurses uniforms using an Internet search-engine, and saw what looked like a comparable uniform, but that would most probably have been true of any nurse's uniform I encountered, and maybe they aren't so dissimilar anyway. I recently read the following comment by Richard Wiseman, a psychologist and ghost-buster from the University of Hertfordshire:

> The majority of these [ghostly] experiences happen in old buildings with a tremendous sense of history, and people are aware of this. Hospitals are inherently places that are associated with death. Nurses in particular have to cope with life and death on a daily basis. At

some level, there is a need to believe in ghosts and an afterlife, as a way of saying death is not final, as a comfort. *The Guardian* 22 December 2004.

Ghostly sightings it seems are common in hospitals, and are most often recorded (not surprisingly) amongst nurses and other medical staff. While my sighting has something in common with this position, there are obvious and significant differences as well.

+++

So, not a classic ghost story by any means, and no great insight into the relations between historic accuracy and the paranormal and inexplicable. Neither does it come close to Edensor's (2005) nor does it come close to Bell's (1997) rigorous treatment of spirit and phenomenology, memory and dereliction. It is merely something that I once experienced and have thought a lot about, on and off, over the years. Thinking about this now, 35 years later, my questions about the significance of this experience are very different to the ones I struggled to form back then. Certainly recalling these events gave a small degree of comfort when my father died in 2001 – perhaps only in recalling the seriousness with which he treated this matter. It would have been so easy to laugh it off, to dismiss it … but he did not. But maybe there is something more in this experience. It is, for example, impossible for me now to dismiss ghosts as something that cannot exist. Hauntings, or at least the potential for hauntings, are ever-present in my explorations of abandoned and empty sites and buildings. Perhaps they do defy logic and reason. But something inexplicable exists. I *know* it does. The fact that my experience came as it did, in a context directly compatible with what was to become a close personal and professional interest, can be no coincidence. I can never visit an abandoned military site now without recalling this experience. Certainly, the details will have been changed over time, the picture diluted or exaggerated; but it happened, and it is still there as I wander into empty rooms on a former airbase or an underground bunker. The benefit of this is that I have no trouble peopling these places, and recalling the significance they once had, and continue to have, for their former residents. The ghosts of place exist, as I hope some of the chapters in this book have demonstrated. The places I have described are not mere physical constructions devoid of human interest and social meaning. They have both of these things in abundance and, for me at least, the ghosts of place are a constant reminder of this.

References

Addison, P. and Crang, J. A. (eds), 2000. *The Burning Blue: A New History of the Battle of Britain*. London: Pimlico

Anderson, J. 2004. Talking whilst walking: a geographical archaeology of knowledge. *Area* 36.3, 254–261

Anderton, M. and Schofield, J. 1999. Anti-aircraft gunsites, then and now. *Conservation Bulletin* 36, 11–13

Angelo, J. A. Jr. and Buden, D. 1985. *Space Nuclear Power*. Malabar, FL: Orbit Book Company, Inc

Anon, 1993. Victory to the Dongas. London Psychogeographic Association Newsletter, Accessible at http://www.unpopular.demon.co.uk/elpan001/elpan001.html)

Anon, 2004. Foreword. In Dolff-Bonekämper G. (ed), *Dividing lines, connecting lines – Europe's cross-border heritage*, 9. Strasbourg: Council of Europe

Anon, 2005. Contemporary art taps vitality of Boston's historic sites. *Common Ground* 10 (3), 4–7

Armitage, J. In press. Military bunkers. In Hutchison, R. (ed), *Encyclopedia of Urban Studies*. London: Sage

Ash, T. G. 1997. *The File: A Personal History*. London: HarperCollins

Association for Studies in the Conservation of Historic Buildings, 2001. *Transactions* 26

Baker, F. 1993. The Berlin Wall: production, preservation and consumption of a 20th Century monument. *Antiquity* 67, 709–733

Baker, D. 1996. *Spaceflight and Rocketry: A Chronology*. New York, NY: Facts on File, Inc

Baker, D. 1999. Introduction: contexts for collaboration and conflict. In Chitty, G. and Baker, D. (eds), *Managing Historic Sites and Buildings: Reconciling Presentation and Preservation*, 1–21. London: English Heritage and Routledge

Ballantyne, R. and Uzzell, D. 1993. Environmental mediation and hot interpretation: a case study of District Six, Cape Town. *Journal of Environmental Education* 24 (3), 4–7

Barrett, J., Bradley, R. and Green, M. 1991. *Landscape, Monuments and Society*. Cambridge: Cambridge University Press

Bartolomé, J. F. and Tipper, M. 2005. Heritage deviations in relation to town planning in Ciutat de Mallorca. *Journal of Mediterranean Studies* 15.2, 427–480

Beazley, O. 2007. A paradox of peace: The Hiroshima Peace Memorial (Genbaku Dome) as world heritage. In Schofield, J. and Cocroft, W. D. (eds), *A Fearsome Heritage: Diverse Legacies of the Cold War*, 33–50. Walnut Creek: Left Coast

Beck, C. M. 2002. The archaeology of scientific experiments at a nuclear testing ground. In Schofield J., Johnson W. G. and Beck C. M. (eds), *Matériel Culture: The Archaeology of Twentieth Century Conflict*, 65–79. London: Routledge

Beck, C. M. and Green, H. L. (eds), 2004. *Al O'Donnell: Oral History*. Las Vegas, NV: Desert Research Institute

Bedford, E. and Murinik, T. nd. Re-membering that place: public projects in District Six. In Soudien, C. and Mayer, R. nd (eds), *The District Six Public Sculpture Project*, 12–22. Cape Town: The District Six Museum Foundation

Beedle, J. 1996. Tangmere. In Ramsey, W. G. (ed), *Battle of Britain,* 30–41. London: After the Battle

Bell, M. 1997. The ghosts of place. *Theory and Society* 26, 813–836

Ben-Ze-ev, E. and Ben-Ari, E. 1996. Imposing politics: failed attempts at creating a museum of 'co-existence' in Jerusalem. *Anthropology Today* 12.6, 7–13

Berger, J. 1972. *Ways of Seeing*. London: British Broadcasting Corporation and Penguin Books

Blackwood, C. 1984. *On the Perimeter*. London: Flamingo

Boissevain, J. 1993. *Saints and Fireworks: religion and politics in rural Malta*. Valletta: Progress

Borneman, J. 1992. *Belonging in the two Berlins: kin, state, nation*. Cambridge: Cambridge University Press

Borneman, J, 1998. *Grenzregime* (border regime): the Wall and its aftermath. In Wilson, T. M. and Donnan, H. (eds), *Border Identities: Nation and state at international frontiers,* 162–190. Cambridge: Cambridge University Press

De Botton, A. 2003 [2002]. *The art of travel*. London: Penguin

Boulton, A. 2001. *Cood bay Forst Zinna*. (Film, privately distributed.)

Boulton, A. 2007. Cood bay Forst Zinna. In Schofield, J. and Cocroft, W. D. (eds), *A Fearsome Heritage: Diverse Legacies of the Cold War*, 181–192. Walnut Creek: Left Coast

Bourne, J. M. 1989. *Britain and the Great War 1914–1918*. New York: Edward Arnold

Bourriaud, N. 2002 [1998]. *Relational Aesthetics*. Dijon: Les Presses du Réel

Bradley, A., Buchli, V., Fairclough, G., Hicks, D., Miller, J. and Schofield J. 2004. *Change and Creation: Historic Landscape Character 1950–2000*, London: English Heritage

Broekman, P van M. and Berry, J. 2002. Countdown to Zero, Count up to Now. Mute Magazine 25 (28 November 2002). Viewed online at http://www.metamute.com/look/article.tpl?IdLang uage=1&IdPublication=1&NrIssue=25&NrSection=10&NrArticle=760

Brown, G. 2001. Listening to queer maps of the city: gay men's narratives of pleasure and danger in London's East End. *Oral History,* Spring, 48–61

Bryant, B. 1995. *Twyford Down – roads, campaigning and environmental law*. London: Taylor and Francis Group

Buchli, V. 2007. Cold war on the domestic front. In Schofield, J. and Cocroft, W. D. (eds), *A Fearsome Heritage: Diverse Legacies of the Cold War*, 211–220. Walnut Creek: Left Coast

Buchli, V. and Lucas, G. 2001. *Archaeologies of the contemporary past*. London and New York: Routledge

Butigan, K. 2003. *Pilgrimage through a Burning World: Spiritual Practice and Nonviolent Protest at the Nevada Test Site*. Albany: State University of New York Press

Butler, B. 1996. The tree, the tower and the shaman: the material culture of resistance of the No M11 Link Roads Protest of Wanstead and Leytonstone, London. *Journal of Material Culture* 1.3, 337–363

Byrne, D. 2008. Heritage as social action. In Fairclough, G., Harrison, R., Jameson, J. Jnr. and Schofield, J. (eds), *The Heritage Reader*, 149–173. London: Routledge

Byrne, D., Brayshaw, H. and Ireland, T. 2001. *Social significance: a discussion paper*, New South Wales National Parks and Wildlife Service

Calder, A. 1969. *The People's War 1939–45*. London: Cape

Cantwell, J. D. 1993. *The Second World War: A guide to documents in the Public Record Office*. London: HMSO

Carman, J. 1997. Approaches to Violence. In J. Carman (ed), *Material Harm: archaeological studies of war and violence* 1–23, Glasgow: Cruithne

Casha, M. and Radmilli, R. 2005. *Il na Beltin (Voices of Valletta)*. Film produced for Med-Voices Project Malta

Cassar, P. 1964. *Medical History of Malta*. London: Wellcome Historical Medical Library

Cave, N. 2000. Battlefield Conservation: First International Workshop in Arras 29 February to 4 March 2000. *Battlefields Review* 10, 41–60

Childers, E. 1903 [1995]. *The Riddle of the Sands*. London: Penguin Popular Classics

Chippindale, C. 1997. Editorial. *Antiquity* 71(273), 505–512

Churchill, W. S. 1949. *The Second World War, Vol.* 2, *Their Finest Hour.* London: Cassell

Clark, K. 2002. In small things remembered: significance and vulnerability in the management of Robben Island World Heritage Site. In Schofield, J., Johnson, W. G. and Beck, C. M. (eds), *Matériel Culture: the archaeology of twentieth century conflict,* 266–280. London and New York: Routledge. One World Archaeology 44

Cocroft, W. D. 2000. *Dangerous Energy: the archaeology of gunpowder and military explosives manufacture.* London: English Heritage

Cocroft, W. 2001. *Cold War Monuments: An Assessment by the Monuments Protection Programme.* London: English Heritage

Cocroft, W. 2003. *Atomic Weapons Research Establishment, Foulness, Essex.* Archaeological Investigation Report Series AI/21/2004. Cambridge: English Heritage

Cocroft, W. 2008. Germany and the public presentation of public rocket sites: a personal view. *Prospero – The Journal of British Rocketry and Nuclear History* 5, 15–33

Cocroft, W. and Thomas, R. J. C. 2003. *Cold War: Building for Nuclear Confrontation 1946–1989.* London: English Heritage

Cocroft, W., Devlin, D., Schofield, J. and Thomas, R. J. C. 2006. *War Art: Murals and Graffiti – Military Life, Power and Subversion.* York, UK: Council for British Archaeology

Cohen-Joppa, J. and Cohen-Joppa F. 1990. Nuclear Resister, 19 January 1990 (http://www.uq.net.au/%7Ezzdkeena/NvT/16/16.9.txt)

Collier, R. 1966. *Eagle Day.* London: Pan.

Corbell, P. 1996. Kenley. In Ramsey, W. G. (ed), *Battle of Britain,* 46–51. London: After the Battle

Corin, L. G. (ed) 1999. *Jane and Louise Wilson: Stasi City Gamma Parliament Las Vegas, Graveyard Time.* London: Serpentine Gallery

Cresswell, T. 2004. *Place: A Short Introduction.* Oxford: Blackwell

Cunningham, C. 1994. *The Beaulieu River goes to war.* Brockenhurst: Montagu Ventures

Darvill, T. and Fulton, A. 1998. *MARS: The Monuments at Risk Survey of England* 1995. London: Bournemouth University and English Heritage

Delgado, J. 1996. *Ghost Fleet: The Sunken Ships of Bikini Atoll.* Honolulu: University of Hawaii Press

Deller, J. nd. *The English Civil War Part II: Personal Accounts of the 1984–1985 Miners' Strike.* Manchester: Artangel

Dench, G. 1975. *Maltese in London: A Case-Study in the Erosion of Ethnic Consciousness.* London and Boston: Routledge and Kegan Paul

Dobinson, C. 1996a. *Twentieth Century Fortifications in England. Vol. 1. Anti-aircraft artillery 1914–1946.* Unpublished report for English Heritage

Dobinson, C. 1996b. *Twentieth Century Fortifications in England, Vol.5: Operation Overlord.* Unpublished report for English Heritage

Dobinson, C. 1996c. *Twentieth-Century Fortifications in England, Vol. 2, Anti-invasion Defences of World War II.* Unpublished report for English Heritage

Dobinson, C. 1997. *Twentieth-Century Fortifications in England, Vol. 9, Airfield Themes.* Unpublished report for English Heritage

Dobinson, C. S. 1998. *Twentieth Century Fortifications in England: Vol. X: The Cold War.* Unpublished report for English Heritage

Dobinson, C. 1999a. *Twentieth Century Fortifications in England. Vol. VI. Coast artillery: England's fixed defences against the warship, 1900–1956.* Unpublished report for English Heritage

Dobinson, C. 1999b. *Twentieth Century Fortifications in England. Vol. VII. Acoustics and radar: England's early warning systems, 1915–1945.* Unpublished report for English Heritage

Dobinson, C. 2000a. *Fields of Deception: Britain's bombing decoys of World War II.* London: Methuen

Dobinson, C. 2000b. *Twentieth Century Fortifications in England. Supplementary study. Experimental and Training Sites: An Annotated Handlist.* Unpublished report for English Heritage

Dobinson, C. 2001. *AA Command: Britain's Anti-Aircraft Defences of the Second World War.* London: Methuen

Dobinson, C. S., Lake, J. and Schofield, A. J. 1997. Monuments of War: defining England's 20th-century defence heritage. *Antiquity* 71, 288–299

Dodds, J. D. 1998. Bridge over the Neretva. *Archaeology* 51.1, 48–53

DoE, 1990. *Planning Policy Guidance Note 16: Archaeology and Planning*. London: HMSO

Dolff-Bonekaemper, G. nd. Sites of historical significance and sites of discord: historic monuments as a tool for discussing conflict in Europe. In *Forward Planning: the function of cultural heritage in a changing Europe*, 53–57. Council of Europe

Dolff-Bonekaemper, G. 2002. The Berlin Wall: an archaeological site in progress. In Schofield, J., Johnson, W. G. and Beck, C. (eds), *Matériel Culture: The Archaeology of Twentieth Century Conflict*, 236–248. London: Routledge: One World Archaeology, 44

Dolff-Bonekämper, G. 2004. Introduction. In Dolff-Bonekämper G. (ed), *Dividing Lines, Connecting Lines – Europe's Cross-Border Heritage*, 11–16. Strasbourg: Council of Europe

Doughty, M. (ed). 1994. *Hampshire and D-Day*. Crediton: Hampshire Books

Dowson, T. 2000. Why queer archaeology? An introduction. *World Archaeology* 32.2, 161–165

Edensor, T. 2005. *Industrial Ruins: Space, Aesthetics and Materiality*. Oxford and New York: Berg

Eksteins, M. 1994. Michelin, Pickfords et La Grande Guerre: Le Tourisme sur Le Front Occidental: 1919–1991. In Becker, J.-J., Winter, J., Becker, A. and Audoin-Rouzeau, S. (eds), *Guerre et Cultures, 1814–1918*, 417–428. Paris: Armand Colin

Elliott, P. 2003. Presenting reality: an introduction to the Boyle Family. In Elliott, P., Hare, B. and Wilson, A. (eds), *Boyle Family*, 9–19. Edinburgh: National Galleries of Scotland

English Heritage 1998. *Monuments of War: The Evaluation, Recording and Management of Twentieth-Century Military Sites*. London: English Heritage

English Heritage 2000. *MPP 2000: A Review of the Monuments Protection Programme, 1986–2000*. London: English Heritage

English Heritage 2008a. *Conservation Principles: Policies and Guidance for the Sustainable Management of the Historic Environment*. London: English Heritage

English Heritage 2008b [1997]. Sustaining the historic environment: new perspectives on the future. In Fairclough, G., Harrison, R., Jameson, J. Jnr. and Schofield, J. (eds), *The Heritage Reader*, 313–321. London: Routledge

Evans, K. 1998. *Copse: The Cartoon Book of Tree Protesting*, Biddestone: Orange Dog Publications

Fairclough, G. 2002. Towards integrated management of a changing landscape: Historic Landscape Characterisation in England. In Swensen, G. (ed), *Cultural Heritage on the Urban Fringe*, 29–39. Nannestad workshop report, NIKU publikasjoner 126

Fairclough, G. 2003. Cultural landscape, sustainability and living with change. In Teutonico, J.-M., and Matero, F. (eds), *Managing Change: Sustainable Approaches to the Conservation of the Built Environment*, 23–46. Los Angeles, CA: The Getty Conservation Institute

Fairclough, G. and Rippon, S. (eds), 2002. *Europe's Cultural Landscape: Archaeologists and the Management of Change*, Europae Archaeologiae Consilium Occasional Paper Number 2

Fairclough, G. and Schofield, J. 2000. *Proposed Scheduling at Twyford Down, to Include a Monument to Environmental Protest*, English Heritage internal committee paper

Farber, L. 2005. Dis-location/Re-location: 'Implanting Africa'. *Cultural Politics* 1.3, 301–312

Feversham, P. and Schmidt, L. 1999. *The Berlin Wall Today*. Berlin: Verlag Bauwesen

Finn, S. 1973. *Lincolnshire at War, 1939–1945*. Brayford

Finn, C. 2004. *Past Poetic: Archaeology and the Poetry of W.B. Yeats and Seamus Heaney*. London: Duckworth

Foot, W. 2006. *Beaches, Fields, Streets and Hills: The Anti-Invasion Landscapes of England*. York: Council for British Archaeology

Fortune, L. 1996. *The House in Tyne Street: Childhood Memories of District Six*. Cape Town: Kwela

Forty, A. 1999. Introduction. In Forty A. and Küchler S. (eds), *The Art of Forgetting*, 1–18. Oxford: Berg

Forty, A. and Küchler, S. (eds), 1999. *The Art of Forgetting*. Oxford: Berg

Francis, P. 1996. *British Military Airfield Architecture*. Somerset, England: Patrick Stephens Ltd

Francis, P. 1997. *United States Naval Amphibious Supply Base Exeter*. Exeter: Airfield Research Publishing

Fransen H. 1996. The architecture that Cape Town lost. In Greshoff J. (ed), *The Last Days of District Six*, 19–21. Cape Town: The District Six Museum Foundation

Funder, A. 2003. *Stasiland: Stories from Behind the Berlin Wall*. London: Granta Books

Gabie, N. 2005. Coast. Colchester: firstsite. Distributed by Cornerhouse Publications (http://www.cornerhouse.org)

Gaddis, J. L. 1997. *We Now Know: Rethinking Cold War History*. Oxford: Clarendon

Galea, D. (prod.) 2004. *Tribute to the Swinging Pioneers of Malta*. Heritage Records, 107

Gane, M. 1999. Paul Virilio's bunker theorizing. *Theory, Culture and Society* 16.5, 85–102

Gane, M. 2007. Ghosts of Place (review of Combat Archaeology). *Cultural Politics* 3.1, 133–136

Gard'ner, J. 2005. Heritage protection and social inclusion: a case study from the Bangladeshi community of East London, *International Journal of Heritage Studies* 10.1, 75–92

Geiryn, T. F. 1998. Balancing acts: science, Enola Gay and history wars at the Smithsonian. In Macdonald S. (ed), *The Politics of Display: Museums, Science, Culture*, 197–228. London and New York: Routledge

Giddens, A. 1991. *Modernity and Self-Identity: Self and Society in the Late Modern Age*. Cambridge: Polity

Gilbert, M. 1998. *Holocaust Journey: Travelling in Search of the Past*. London: Phoenix

Gilchrist, R. (ed) 2003. *The Social Commemoration of Warfare*. World Archaeology Vol. 35.1, 79–97

Gilens, A. 1995. *Discovery and Despair: Dimensions of Dora*. Berlin/Bonn: Westkreuz-Verlag

Gilfoyle, T. J. 2005. Archaeologists in the brothel: 'Sin City', historical archaeology and prostitution. *Historical Archaeology* 39.1, 133–141

Goldberg, R. L. 1979. *Performance Art, from Futurism to the Present*. London: Thames and Hudson

Graves Brown, P. 1996. Road to Nowhere? *Museums Journal* 96(11), 25–27

Graves Brown, P. (ed), 2000. *Matter, Materiality and Modern Culture*. London and New York: Routledge

Hall, M. 1998. Cape Town's District Six and the Archaeology of Memory. Precirculated paper presented to the World Archaeology Congress's Inter-Congress on damage to cultural property, Croatia

Hall, M. 2000. *Archaeology and the Modern World: Colonial Transcripts in South Africa and the Chesapeake*. London and New York: Routledge

Hall, M. 2001. Social archaeology and the theatres of memory. *Journal of Social Archaeology* 1.1, 50–61

Halle, L. J. 1967. *The Cold War as History*. New York: Harper and Row Publishers

Harris, A. 1994. Gathering Mulberries, *History Today* (*June* 1994), 15–22

Harwit, M. 1996. *An Exhibit Denied: Lobbying the History of Enola Gay*. New York, NY, Copernicus

Hastings, M. 1976. *Bomber Command*. London: Michael Joseph

Hawkes, C. F. C., Myers, J. N. L. and Stevens, M. A. 1930. Saint Catherine's Hill, Winchester. *Proceedings of the Hampshire Field Club and Archaeological Society* 11, 1–286

Hayden, D. 1995. *The Power of Place: Urban Landscapes as Public History*. Cambridge, MA: MIT

Haysom, D. 2002. Special Signals Unit History Berlin Part 4. *Rafling News*, September 2002. No page references

Hebdige, D. 1992. Digging for Britain: An excavation in seven parts. In Strinati, I. and Wagg, S. (eds), *Come on Down: Popular Media Culture in Post-War Britain,* 325–377. London: Routledge

Hems, A. 2006. Introduction: beyond the graveyard – extending audiences, enhancing understanding. In Hems A. and Blockley, M. (eds), *Heritage Interpretation*, 1–7. London: English Heritage and Routledge.

Hennessy, P. 1992. *Never Again: Britain, 1945–1951*. London

Hicks, D. and Beaudry, M. (eds) 2006. *The Cambridge Companion to Historical Archaeology*. Cambridge: Cambridge University Press

Hillary, R. 1997 [1942]. *The Last Enemy*. London: Pimlico

Hinchliffe, R. 1997. The Cold War: the need to remember or desire to forget. *History Workshop Journal* 43, 234–239

Hipperson, S. 1998. Greenham Common. In MacArthur, B. (ed), *The Penguin Book of Twentieth Century Protest*, 355–357. London: Viking

Hoare, P. 2001. *Spike Island: The Memory of a Military Hospital*. London: Fourth Estate

Hobsbawm, E. 1995. *Age of Extremes: The Short Twentieth Century, 1914–91*. London: Abacus

Holmes, R. 1997. *War Walks 2: From the Battle of Hastings to the Blitz*. London: BBC Books

Holtorf, C. 2004a. Review of *Matériel Culture*. *Fornvännen* 99: 317–318

Holtorf, C. 2004b. Incavation – Excavation – Exhibition. In Brodie, N. and Hills, C. (eds), *Material engagements: Studies in Honour of Colin Renfrew*, 45–53. Cambridge: MacDonald Institute Monographs

Holtorf, C. 2005. *From Stonehenge to Las Vegas: Archaeology as Popular Culture*. Oxford: AltaMira

Holyoak, V. 2001. Chalk military badges on Fovant Down. English Heritage Schedule Entry

Honnef, K., Sieloff, M. and Strauss, J. 1998. *Giestesstatt: Konversionskunst in W nsdorf-Waldstadt 1997*. Potsdam: J. Strauss Verlag

Horn, M. 2000. The concept of Total War: national effort and taxation in Britain and France during the First World War. *War and Society* 18.1, 1–22

Howard, P. 1991. *Landscapes: The Artist's Vision*. London: Routledge

Hughes, M. 1994. D-Day archaeology. In Doughty, M. (ed), *Hampshire and D-Day*, 162–168. Hampshire Books: Crediton

Hunter, J. and Ralston, I. (eds), 2007. *Archaeological Resource Management in the UK: An introduction (second edition)*. Stroud: Sutton

Ioffe, G. and Netedova, T. 2001. Land-use changes in the environs of Moscow. *Area* 33.3, 269–276

James, N. 2002. The Cold War. *Antiquity* 76, 664–666

Jameson, J. H. Jnr., Ehrenhard, J. E. and Finn, C. A. (eds) 2003. *Ancient Muses: Archaeology and the Arts*. Tuscaloosa and London: University of Alabama Press

Jarman, N. 2002. Troubling remnants: dealing with the remains of conflict in Northern Ireland. In Schofield, J., Johnson, W. G. and Beck, C. M. (eds), *Matériel Culture: The Archaeology of Twentieth Century Conflict*, 281–295. London and New York: Routledge. One World Archaeology 44

Jeppie S. and Soudien, C. (eds) 1990. *The Struggle for District Six: Past and Present*. Cape Town: Buchu Books

Johnson, W. G. 2002. Archaeological examination of Cold War architecture: a reactionary cultural response to the threat of nuclear war. In Schofield, J., Johnson, W. G. and Beck, C. M. (eds), *Matériel Culture: The Archaeology of Twentieth Century Conflict*, 227–235, London and New York: Routledge. One World Archaeology 44

Johnson, W. G. and Beck, C. 1995. Proving Ground of the Nuclear Age. *Archaeology* 48.3, 43–49

Jones, D. (ed), 2002. *20ᵗʰ Century Heritage: Our Recent Cultural Legacy*. Proceedings of the Australia ICOMOS National Conference 2001, School of Architecture, Landscape Architecture and Urban Design, The University of Adelaide, and Australia ICOMOS

July Skies 2004. *The English Cold* (audio CD). Make Mine Music

Junor, B. 1995. *Greenham Common Womens Peace Camp: A History of Non-Violent Resistance 1984–1995*. London: Working

Kenney, M. R. 1998. Remember, Stonewall was a riot: understanding gay and lesbian experience in the city. In Sandercock, L. (ed), *Making the Invisible Visible: A Multicultural Planning History*, 120–132., Berkeley, CA, University of California Press

King, A. 1998. *Memorials of the Great War in Britain. The Symbolism and Politics of Remembrance*. Oxford and New York: Berg

King, T. F. 2003. *Places that Count: Traditional Cultural Properties in Cultural Resource Management*. Walnut Creek: Altamira

Kippin, J. 2001. *Cold War Pastoral: Greenham Common*. London: Black Dog Publishing

Kruckewitt, J. 2001.*The Death of Ben Linder: The Story of a North American in Sandinista Nicaragua*. New York: Seven Stories

Kuipers, M. and Schofield, J. 2004. Lines of tension. In Dolff-Bonekämper G. (ed), *Dividing Lines, Connecting Lines – Europe's Cross-Border Heritage*, 29–47. Strasbourg: Council of Europe

Kuletz, V. 1998. *The Tainted Desert: Environmental and Social Ruin in the American West*. New York and London: Routledge

Kushner, T. (ed) 1992. *The Jewish Heritage in British History: Englishness and Jewishness*. London: Frank Cass and Co., Ltd

Kyriakides, Y. 2001. a conSPIracy cantata (audio CD). Amsterdam: unsounds (http://www. unsounds.com)

Kyriakides, Y. 2007. Voices in limbo: *a conspiracy cantata and the Buffer Zone*. In Schofield, J. and Cocroft, W. D. (eds), *A Fearsome Heritage: Diverse Legacies of the Cold War*, 221–238. Walnut Creek: Left Coast

Ladd, B. 1997. The Ghosts of Berlin: *Confronting German history in the urban landscape*. Chicago and London: University of Chicago Press.

Ladd, B. 2002. East Berlin political monuments in the late German Democratic Republic: finding a place for Marx and Engels. *Journal of Contemporary History* 37.1, 91–104

Lake, J. 2000. Survey of Military Aviation Sites and Structures: Summary Report. Unpublished report by Thematic Listing Programme, English Heritage

Lake, J., Dobinson, C. and Francis, P. 2005. The evaluation of military aviation sites and structures in England. In Hawkins, B., Lechner, G. and Smith, P. (eds), *Historic Airports. Proceedings of the International 'L'Europe de l'Air' Conferences on Aviation Architecture*. London: English Heritage

Langlands, B. and Bell, N. 2004. *The house of Osama Bin Laden*. London: Thames and Hudson

Layton, R., Stone, P., Thomas, J. and Roa, N. (eds) 2001. *Destruction and Restoration of Cultural Property*. London and New York: Routledge. One World Archaeology 41

le Grange L. 1996. The Urbanism of District Six. In Greshoff, J. *The Last Days of District Six*, 7–15. Cape Town: The District Six Museum Foundation

Linenthal, E. T. 1995. Struggling with history and memory. *Journal of American History* 82, 1094–1101

Loeber, C. R. 2002. *Building the Bombs: A History of the Nuclear Weapons Complex*. Albuquerque, New Mexico: Sandia National Laboratories

London III, J. R. 1993. The preservation of space-related historic sites. *Journal of the British Interplanetary Society* 46, 279–285

Lowenthal, D. 1996. *The Heritage Crusade and the Spoils of History*. London: Viking

Lowenthal, D. 1998. Fabricating Heritage. *History and Memory* 10.1, 5–24

Ludtke, A. 1997. Histories of mourning: flowers and stones for the war dead, confusion for the living – vignettes from East and West Germany. In Sider, G. and Smith, G (eds), *Between History and Histories: The Making of Silences and Commemorations,* 149–179. Toronto: University of Toronto Press

Magro Conti, J. and Darmanin, D. 2003. Scheduled Property Monitoring, Timber Shopfronts and Kiosks in Valletta (1995). Draft unpublished internal report. Malta: Malta Environment and Planning Authority

Malan, A. and Soudien, C. 2002. Managing heritage in District Six, Cape Town: conflicts past and present. In Schofield, J., Johnson, W. G. and Beck, C. M. (eds), *Matériel Culture: The Archaeology of Twentieth Century Conflict*, 249–265. London and New York: Routledge: One World Archaeology 44

Malan, A. and van Heyningen, E. 2001. Twice removed: Horstley Street in Cape Town's District Six, 1865–1982. In Mayne, A. and Murray, T. (eds), *The Archaeology of Urban Landscapes: Explorations in Slumland*, 39–56. Cambridge: Cambridge University Press

Marceau, T. E., Harvey D. W., Stapp D. C., Cannon S. D., Conway C. A., Deford D. H., Freer B. J., Gerber M. S., Keating J. K., Noonan C. F. and Weisskopf G. 2003. *Hanford Site Historic District History of the Plutonium Production Facilities 1943–1990*. Columbus, OH: Batelle

McArthur, B. (ed), 1998. *The Penguin Book of Twentieth Century Protest*. London: Viking

McCrum, R. 1997. Goodbyee? *The Observer* (9/11/97)

McCrystal, C. and Higgins, A. 1998. How (and where) the Russians planned to annihilate you. *The Observer* (10/5/98)

McDonald, K. 1989. *Dive Sussex*. Middlesex, London: Underwater World Publications

McDonald, K. 1994. *Dive Kent*. Middlesex, London: Underwater World Publications

McLachlan, I. 1989. *Final Flights: Dramatic Wartime Incidents Revealed by Aviation Archaeology*. Sparkford: Patrick Stephens Ltd

McOmish, D., Field, D. and Brown, G. 2002. *The Field Archaeology of the Salisbury Plain Training Area*. London: English Heritage

Mills, S. 2005. Applying auditory archaeology to Historic Landscape Characterisation: a pilot project in the former mining landscape of Geevor and Levant Mines, West Penwith, Cornwall. Unpublished report for English Heritage. Available at http://www.english-heritage.org.uk/characterisation

Ministry of Environment 1994. *Fortification in Denmark 1858–1945: A status report (an English unillustrated version of Befæstningsanlæg i Danmark 1858–1945: En statusrapport*, Copenhagen: The Ministry of Environment

Mitchell, J. 2002. *Ambivalent Europeans: Ritual, Memory and the Public Sphere in Malta*. London: Routledge

Moon, G. and Atkinson, R. 1997. Ethnicity. In Pacione, M. (ed), *Britain's Cities: Geographies of Division in Urban Britain*, 262–276. London: Routledge

Moran, L., Skeggs, B., Tyrer, P. and Corteen, K., 2003. The formation of fear in gay space: the 'straights' story, *Capital and Class* 80, 183–198

Morris, R. 1998. Amateurs all the way. *Defence Lines* 11, 7–9

Morris, R. and Dobinson, C. 1995. *Guy Gibson*. London: Penguin Books

Murch. M., Murch, D. and Fairweather, L. 1984. *The American Forces at Salcombe and Slapton During World War Two*. Plymouth: PDS Printers

Nash, F. 1998. *Anti-Invasion Defences in Essex*. Essex County Council report

Newman, D. and Paasi, A., 1998. Fences and neighbours in the postmodern world: boundary narratives in political geography, *Progress in Human Geography* 22.2, 186–207

Nicoll, W. 1996. North Weald. In Ramsey, W. G. (ed), *Battle of Britain*, 160–175. London: After the Battle

Noakes, L. 1997. Making Histories: Experiencing the Blitz in London's Museums in the 1990s. In Evans M. and Lunn K. (eds), *War and Memory in the Twentieth Century*, 89–104. Oxford and New York: Berg

Norris, P. 1996. Northolt. In Ramsey, W. G. (ed), *Battle of Britain,* 236–249. London: After the Battle

Odom, W. E. 1998. *The Collapse of the Soviet Military*. New Haven, CT, Yale University Press

Pacione, M. (ed) 1997. *Britain's Cities: Geographies of Division in Urban Britain*. London: Routledge

Parr, C. 2006. Public art: its role as a medium for interpretation. In Hems, A. and Blockley, M. (eds), *Heritage Interpretation*, 123–140. London: English Heritage and Routledge

Patrizio, A. 2002. *Stefan Gec*. London: Black Dog Publishing Ltd

Peach, P. and O'Brien, M. 2000. Nevada Test Site peacemakers ring in new millennium. *Salt Online* (http://www.bvmcong.org/salt/salt/spring2000/peachobrien.htm)

Peacock, A. J. 1994. Introduction. In *Illustrated Michelin Guides to the Battlefields 1914–1918: The Somme Vols 1–2* (Facsimile copy). York, UK: G. H. Smith and Son Ltd

Pearson, M. and Shanks, M. 2001. *Theatre/Archaeology*. London: Routledge

Pecacz, J. 1994. Did rock smash the Wall? The role of rock in political transition. *Popular Music* 13.l, 41–49

Peckham, I. 1994. *Southampton and D-Day*. Southampton: Southampton City Council

Pelling, H. 1970. *Britain and the Second World War*. Glasgow: Collins

Penrose, S. (with contributors) 2007. *Images of Change: An Archaeology of England's Contemporary Landscape*. London: English Heritage

Perkins, G. 1999. Museum War Exhibits: Propaganda or Interpretation? *Interpretation* 4, 38–42

Pile, F. 1949. *Ack-ack. Britain's Defences Against Air Attack During the Second World War.* London: Harrap

Pinto, R., Bourriaud, N. and Diamianovic, M. 2003. *Lucy Orta.* London: Phaidon Press Ltd

Planel, P. 1995. *A Teacher's Guide to Battlefields, Defence, Conflict and Warfare.* London: English Heritage

Porteous, J. D. 1996. *Environmental Aesthetics: Ideas, Politics and Planning.* London and New York: Routledge

Porteous, J. and Mastin, J. 1985. Soundscape. *Journal of Architectural and Planning Research* 2, 169–186

Pritchard, M. and McDonald, K. 1991. *Dive Wight and Hampshire.* Middlesex, London: Underwater World Publications

Raby, A. nd. Duxford Airfield: The Story of a Famous Fighter Station. Unpublished account

Raby, A. 1996. Duxford. In Ramsey, W. G. (ed), *Battle of Britain*, 198–211. London: After the Battle

Radmilli, R. and Selwyn, T. 2005. Introduction: turning back to the Mediterranean – anthropological issues and the med-voices project. *Journal of Mediterranean Studies* 15.2, 195–218

Ramsey, W. G. (ed), 1978. *Airfields of the Eighth.* London; After the Battle

Ramsey, W. G. (ed), 1996. *The Battle of Britain: Then and Now* (5th ed.). London: After the Battle

Ramsey, W. G. 1997. Memorial to the London Blitz. *After the Battle* 96, 20–25

Raschka, M. 1996. Beirut Digs Out. *Archaeology* 49.4, 44–50

Read, P. 1996. *Returning to Nothing: The Meaning of Lost Places.* Cambridge: Cambridge University Press

Read, P. 2003. *Haunted Earth.* NSW, Australia: University of New South Wales Press

Reed, C. 2003. We're from Oz: marking ethnic and sexual identity in Chicago. *Environment and Planning D: Society and Space* 21, 425–440

Reich, W. 1972. *The Mass Psychology of Fascism.* London: Souvenir

Renfrew, C. 2003. *Figuring it Out: The Parallel Visions of Artists and Archaeologists.* London: Thames and Hudson

Richardson, H. (ed) 1998. *English Hospitals 1660–1948. A Survey of their Architecture and Design.* London: Royal Commission on the Historical Monuments of England

Riley, H. and Wilson-North, R. 2001. *The Field Archaeology of Exmoor.* London: English Heritage

Roberston, M. and Schofield, J. 2000. Monuments in wartime: conservation policy in practice, 1939–45. *Conservation Bulletin* 37, 16–19

Royal Commission on the Historical Monuments of England (RCHME), 1995. Historic Building Report on RAF Bawdsey. Unpublished report

Rupersinghe, K. 1998. *Civil Wars, Civil Peace: An Introduction to Conflict Resolution.* London and Sterling, Virginia: Pluto

Rurrup, R. (ed), 1998. *Topography of Terror: Gestapo, SS and Reiehssicherheitshauptamt on the Prinz Albrecht Terrain.* Berlin

Saint, A. 1996. How Listing Happened. In Hunter, M. (ed), *Preserving the Past,* 115–134. London

Samuel, R. 1994. *Theatres of Memory. Volume 1: Past and Present in Contemporary Culture.* London: Verso

Sassoon, S. 1983. The Aftermath. In *The War Poems of Siegfried Sassoon,* 143. London: Faber and Faber

Saunders, A. 1989. *Fortress Britain: artillery fortification in the British Isles and Ireland.* Beaufort

Saunders, N. J. 2002. The ironic 'culture of shells' in the Great War and beyond. In Schofield, J. Johnson, W. G. and Beck, C. M. (eds), *Matériel Culture: The Archaeology of Twentieth Century Conflict,* 22–40. London and New York: Routledge. One World Archaeology 44

Saunders, N. 2004a. Material culture and conflict: the Great War, 1914–2003. In Saunders, N. (ed), *Matters of Conflict: Material Culture, Memory and the First World War,* 5–25. London: Routledge

Saunders, N. 2004b. Introduction. In Saunders, N. (ed), *Matters of Conflict: Material Culture, Memory and the First World War*, 1–4. London: Routledge

Saxon, E. 1965. The street that shames Hero Island. *Titbits*, February 1965

Schmidt, L. 2005. The Berlin Wall: A Landscape of Memory. In Schmidt, L. and von Preuschen, H. (eds), *On Both Sides of the Wall: Preserving Monuments and Sites of the Cold War Era*, 11–17. Berlin: Westkreuz-Verlag

Schofield, J. 1998. Character, conflict and atrocity: touchstones in the landscapes of war. In Jones M. and Rotherham I. D. (eds), *Landscapes – Perception, Recognition and Management: reconciling the impossible?*, 99–104. Landscape Archaeology and Ecology Vol. 3. Sheffield, UK: The Landscape Forum and Sheffield Hallam University

Schofield, J. 2000. Never mind the relevance: popular culture for archaeologists. In Graves Brown, P. (ed), *Matter, Materiality and Modern Culture*, 131–155. London: Routledge

Schofield, J. 2002a. The role of aerial photographs in national strategic programmes: assessing recent military sites in England. In Bewley, R. H. and Raczkowski, W., (eds), *Aerial Archaeology: developing future practice*, 269–282. IOS: Amsterdam, in co-operation with NATO Scientific Affairs Division

Schofield, J. 2002b. The archaeology of 20th century warfare: global, national and local perspectives. In Jones, D. (ed), *20th Century Heritage – Our Recent Cultural Legacy*, 67–76. School of Architecture, Landscape Architecture and Urban Design, University of Adelaide and Australia ICOMOS Secretariat

Schofield, J. 2003. Memories and monuments in Berlin: a Cold War narrative. *Historic Environment* 17.1, 36–41

Schofield, J. 2005a. *Combat Archaeology: Material Culture and Modern Conflict*. London: Duckworth. Duckworth Debates in Archaeology series

Schofield, J. 2005b. Why write off graffiti? *British Archaeology* 81, 39

Schofield, J. and Cocroft, W. D. (eds), 2007. *A Fearsome Heritage: Diverse Legacies of the Cold War*. Walnut Creek: Left Coast

Schofield, J. and Morrissey, E. 2007. *Titbits* revisited: towards a respectable archaeology of Strait Street, Valletta (Malta). In McAtackney, L., Palus, M. and Piccini, A. (eds) *Contemporary and Historical Archaeology in Theory*, 89–99. British Archaeological Reports

Schofield, J., Johnson, W. G. and Beck, C. M. (eds) 2002. *Matériel Culture: The Archaeology of Twentieth Century Conflict*. London and New York: Routledge. One World Archaeology 44

Scott, E. 1997. Introduction: On the Incompleteness of archaeological narratives. In Moore, J. and Scott, E. (eds) 1997. *Invisible People and Processes – Writing Gender and Childhood into European Archaeology*, 1–12. Leicester: Leicester University Press

Searle, A. 1995. *PLUTO: Pipe-Line Under The Ocean*. Shanklin: Shanklin Chine

Skelton, T. and Valentine, G. (eds) 1998. *Cool Places: Geographies of Youth Culture*. London: Routledge

Seawright, P. 2003. *hidden*. London: The Imperial War Museum

Smith, V. T. C. 1985. Chatham and London: the changing face of English land fortification, 1870–1918. *Post-Medieval Archaeology* 19, 105–149

Smith, D. J. 1989. *Britain's Military Airfields*. Wellingborough

Smith, C. 1997. *Robben Island*. Mayibuye History and Literature Series No. 76. Cape Town: Struik Publishers

Smith, F. M. 1998. Between East and West: sites of resistance in East German youth cultures. In Skelton, T. and Valentine, G. (eds), *Cool Places: Geographies of Youth Culture*, 289–304. London: Routledge

Sockett, E. W. 1991. Stockton-on-Tees 'Y' Station. *Fortress* 8, 51–60

Soudien, C., and Meyer, R. (eds), nd. *The District Six Public Sculpture Project*. Cape Town: The District Six Museum Foundation

Stathatos, J. 2003. Hidding in the Open (sic). In Seawright, P. *hidden*, unpaginated. London: The Imperial War Museum

Stocker, D. 1992. The Shadow of the General's Armchair. *The Archaeological Journal* 149, 415–420

Stuart, J. D. M. and Birkbeck, J. M. 1936. A Celtic Village on Twyford Down – excavated 1933–1934. *Proceedings of the Hampshire Field Club and Archaeological Society* 13, 188–212

Sutton, H. T. 1996. Hornchurch. In Ramsey, W. G. (ed), *Battle of Britain,* 76–87. London: After the Battle

Symmons Roberts, M. 2001. *Burning Babylon.* London: Cape Poetry

Szpanowski 2002. Before and after The Change: the socioeconomic transition period and its impact on the agriculture and cultural landscape of Poland. In Fairclough, G. and Rippon, S. (eds), *Europe's Cultural Landscape: Archaeologists and the Management of Change,* 125–132. EAC Occasional Paper 2

Tarlow, S. 1997. An archaeology of remembering: death, bereavement and the First World War. *Cambridge Archaeological Journal* 7.1, 105–121

Taylor, R. K. S. and Pritchard, C. 1980. *The Protest Makers: The British Nuclear Disarmament Movement of 1958–1965, Twenty Years On.* Oxford: Pergamon

The Ecclesiological Society, 1994. *The Past, Present and Future of St Ethelburga Bishopsgate, Partially Destroyed by Bomb, April 24th 1993.* London

The Ludlow Collective 2001. Archaeology of the Colorado Coal Field War 1913–1914. In Buchli, V. and Lucas, G. (eds), *Archaeologies of the Contemporary Past,* 94–107. London: Routledge

Thomas, M. 2001. *A Multicultural Landscape: National Parks and the Macedonian Experience.* Australia: New South Wales National Parks and Wildlife Service and Pluto

Thomas, M. 2002. *Moving Landscapes: National Parks and the Vietnamese Experience.* Australia: New South Wales National Parks and Wildlife Service and Pluto,

Tilley, C. 1994. *A Phenomenology of Landscape.* Oxford: Berg

Toyofumi, O. 1994. *The Atomic Bomb and Hiroshima.* Tokyo, Japan: Liber

Traynor, I. 1997. Storm breaks over Hitler's eyrie. *The Guardian* (24/11/97)

Trillin, C. 1994. Drawing the line. *The New Yorker* (12/12/94), 50–62

Triscott, N. and La Frenais, R. (eds), 2005. Zerogravity: A Cultural User's Guide. Belgium: The Arts Catalyst. (http://www.artscatalyst.org)

Tschumi, B. 2000. Foreword. In Virilio, P., *A Landscape of Events,* viii–ix. Cambridge, MA: MIT

Turner, S. 2006. *Seafort.* Ramsgate: The Seafort Project

Uzzell, D. 1989. The Hot Interpretation of War and Conflict. In Uzzell D. (ed), *Heritage Interpretation Vol. 1: The Natural and Built Environment,* 33–47. London: Belhaven

Uzzell, D. 1998. The hot interpretation of the Cold War. In English Heritage, *Monuments of War: The Evaluation, Recording and Management of Twentieth-Century Military Sites,* 18–21. London: English Heritage

Uzzell, D. and Ballantyne, R. 2008 [1999]. Heritage that Hurts: interpretation in a postmodern world. In Fairclough, G., Harrison, R., Jameson, J. Jnr. and Schofield, J. (eds), *The Heritage Reader,* 502–513. London: Routledge

Vanderbilt, T. 2002. *Survival City: Adventures Among the Ruins of Atomic America.* New York: Princeton Architectural Press

Virilio, P. 1994. *Bunker Archeology.* Paris: Les Editions du Semi-Circle. (Trans. From the French by George Collins.)

Virilio, P. 2000. *A Landscape of Events* (Translated by Julie Rose). Cambridge, MA: MIT

Virilio, P. 2002. *Desert Screen: war at the speed of light* (trans. from the French by Michael Degener). London and New York: Continuum

Virilio, P. and Lotringer, S. 1997. *Pure War (Revised Edition).* Semiotext(e) Foreign Agents Series

Wainwright, A. 1996. Orford Ness. In Morgan Evans, D., Salway, P. and Thackray, D. (eds), *The Remains of Distant Times: Archaeology and the National Trust,* 198–210. Boydell

Walker, K. E. and Farwell, D. E. 2000. *Twyford Down, Hampshire. Archaeological Investigations on the M3 Motorway from Bar End to Compton, 1990–1993,* Hampshire Field Club Monograph 9 and Wessex Archaeology

Wallace, G. 1975 [1957]. *RAF Biggin Hill.* London

Walley, F. 2001. From bomb shelters to post-war buildings: 40 years' work as a civil engineer in government. *The Structural Engineer* 79.4, 15–21

Watson, F. 2004. The Hush House: Cold War Sites in England. London: Hush House Publishers (http://www.thehushhouse.com)

Watson, F. 2007. The noise of war, the silence of the photograph. In Schofield, J. and Cocroft, W. D. (eds), *A Fearsome Heritage: Diverse Legacies of the Cold War*, 239–252. Walnut Creek: Left Coast

Webster, D. 1997. *Aftermath: The Remnants of War*. London: Constable

Weight, A. 2004. The aftermath of September 11 and the War in Afghanistan. In Langlands, B. and Bell, N. *The House of Osama Bin Laden*, 284–286. London: Thames and Hudson

Weinberg, J. and Elieli, R. 1995. *The Holocaust Museum in Washington*. New York: Rizzoli International Publications

Welch, C. 1997. An investigation of a possible trench 'model' on the site of the First World War camp at Rugeley. Staffordshire County Council Environmental Planning Unit, Research Report No. 2

Wharton, J. 1999. Interpreting the Cold War. *CRM* 22.9, 47–48

Wharton, M. 2002. Evaluating and managing Cold War era historic properties: the cultural significance of US Air Force defensive radar systems. In Schofield, J., Johnson, W. G., and Beck, C. M. (eds), *Matériel Culture: The Archaeology of Twentieth Century Conflict*, 216–226. London and New York: Routledge. One World Archaeology 44

Williams, T. T. 1990. The clan of one-breasted women, *Northern Lights* 6.1, 9–11

Williams, T. T. 2002. A Rock of Resistance. (http://www.arockofresistance.org)

Wills, H. 1985. *Pillboxes: A Study of UK Defences 1940*. Leo Cooper in association with Secker and Warburg

Wills, H. 1994. Archaeological aspects of D-Day: Operation Overlord, *Antiquity* 68, 843–845

Wilson, E. O. 2003a. *Consilience: The Unity of Knowledge*. London: Little, Brown and Co

Wilson, L. K. 2003b. *Spadeadam*. (Film, privately distributed)

Wilson, L. K. 2006a. Out to the Waste: Spadeadam and the Cold War. In Schofield, J. and Cocroft, W. D. (eds), *A Fearsome Heritage: The Diverse Legacies of the Cold War*, 155–180. Walnut Creek: Left Coast

Wilson, L. K. 2006b. *A Record of Fear*. Exhibition catalogue

Wilson, T. M. and Donnan, H. 1998. Nation, state and identity at international borders. In Wilson, T. M. and Donnan, H. (eds), *Border Identities: Nation and State at International Frontiers*, 1–30. Cambridge: Cambridge University Press

Winter, J. 1995. *Sites of Memory; Sites of Mourning*. Cambridge: Cambridge University Press

Wood, D. 1996. *Attack Warning Red: The Royal Observer Corps and the Defence of Britain, 1925–1992*. London: Macdonald and Jane's

Woodward, Sir L. 1967. *Great Britain and the War of 1914–1918*. London: Methuen

Zoviet*france 2000. *The Decriminalisation of Country Music: Themes for Tramway* (audio CD TRAM 1)

Index